POWER-LINED

Electricity, Landscape,
and the American Mind

DANIEL L. WUEBBEN

University of Nebraska Press | Lincoln

Library of Congress
Cataloging-in-Publication Data
Names: Wuebben, Daniel L., author.
Title: Power-lined: electricity,
landscape, and the American
mind / Daniel L. Wuebben.
Description: Lincoln: University of
Nebraska Press, [2019] | Includes bibli-
ographical references and index.
Identifiers: LCCN 2018048123
ISBN 9781496203663 (cloth: alk. paper)
ISBN 9781496215963 (epub)
ISBN 9781496215970 (mobi)
ISBN 9781496215987 (pdf)
Subjects: LCSH: Electrification—United
States—History. | Electric lines—
United States—History. | Electric power
distribution—United States—
History. | Landscape changes—
United States—History.
Classification: LCC HD9685.U5 W84
2019 | DDC 333.793/20973—
dc23 LC record available at
https://lccn.loc.gov/2018048123

Set in Minion Pro by E. Cuddy.
Designed by L. Auten.

This book is dedicated to the memories of my mom and my brothers:

Colleen, Jeremiah, and Rob.

Contents

Illustrations

Preface

Two endless parallel lines in swift longitudinal motion . . .
occupied the whole field. There were present no dreams or visions
in any way connected with human affairs, no ideas or impressions
akin to anything in past experience, no emotions, of course no idea
of personality. There was no conception as to what being it was
that was regarding the two lines, or that there existed any
such thing as such a being; the lines and waves were all.

—Qtd. in William James, *The Principles of Psychology*, 1890

PLAY. When I was five years old, I began to play an imaginary video game called "power lines." I played power lines as the family station wagon glided along the tree-lined, four-lane avenues of Omaha, Nebraska, and as it zipped down bucolic two-lane highways. The game's invisible avatar, my "guy," had short black hair, blue jeans, sturdy legs, and lightning-fast steps. The wagon's movement initiated the action: I locked eyes on the line, mentally pressed PLAY, and guy materialized. Guy made simple, acrobatic maneuvers. He could "run" on top of the lines, "jump" sideways between parallel lines, and "super jump!" over the tops of wooden poles, steel tubes, or lattice towers. In those early years the make-believe game was likely inspired by the undulating actions of digital avatars in real video games like *Super Mario Bros.*, *Sonic the Hedgehog*, and *Contra*. I did not memorize levels nor commit controller buttons to muscle memory, but hours of watching my friends play these games sank the digital rhythms into my brain. Similar patterns were then projected onto the lines in the landscape.

Playing my self-invented power lines game satisfied some nebulous desire for order. The spindly lines and jumping courses could be traced, plotted, and rearranged like the green, pink, and yellow ceramic tiles on the bathroom floor or the wooden beams on the church ceiling. Other shapes and movements focused my attention and brightened my eyes—the leather basketball's spinning ribs splashing through a white nylon net is also fixed in my memory. Yet I found myself routinely hypnotized, staring through the window, fixated on the lines that seemed to gallop above the traffic. Looking back, my experience with these real lines resembled an ether-induced vision reported by a Dr. Shoemaker of Philadelphia and repeated by William James: "There was no conception as to what being it was that was regarding the two lines, or that there existed any such thing as such a being; the lines and waves were all."[1] For miles, it seemed, I disappeared; the power lines were all. Sometimes guy followed me, as if the original game was reengineered everywhere we happened to travel.

Into my teens I habitually followed the syncopated swoops across the Great Plains. In these agricultural regions power lines, water towers, and grain silos often visually dominate rolling farmlands and pastures. In my twenties I toured the continent by car, bus, and motorcycle. Romance and randomness inspired most of these road trips, and my mental maneuverings of the black-haired guy faded. Nevertheless, power lines frequently drew my attention roadside. In Utah I registered the sinewy sines and cosines shimmering through the salt flats; I popped the throttle in Montana, and, with a faint beat pounding inside my helmet, I marched in step with anthropomorphic, transformers-like towers; as the Greyhound crossed Kentucky, I let my sight get stitched by the electric looms hanging over the rectangular horse fences; on the Jersey Turnpike I felt lost amid endless electric spools and faceless faux totem poles lined up near the Thomas Edison Service Plaza. The travel modes, backdrops, and systems of imagery constantly changed—the optical attraction to those rippling forms on the fringes remained.

My youthful play of the power lines game came flooding back in the autumn of 2006. I was sitting in the Main Reading Room of the New York Public Library reading Jonathan Edward's "Spider Letter." In 1723, at the age of twenty, Edwards composed a scientific account of how spiders loosened draglines from their bodies and used them to float across the canopy of Connecticut. Edwards's letter includes hand-drawn diagrams detailing the spiders' ingenious combination of silk threads, gentle breeze, leverage, and buoyancy. Recently, scientists found that spiders can detect and respond to electric fields in the atmosphere and that they use electrostatic forces to help them fly, or "balloon," over significant distances.[2] Edwards was excited by flying behaviors he carefully observed and documented, but he was unaware of their actual electric potential. For him the magnificent "spiders that make those curious, network, polygonal webs" further proved the wisdom of the Creator's design.[3] While sunlight from Bryant Park poured through arched windows into the Reading Room, I read about a teenage Puritan adjusting his eyes to sunbeams reflecting spider silk in the forest.

PAUSE. I leaned back in the wooden chair, looked up, and watched an invisible thread swing from a chandelier's exposed bulb to a golden plaster rosette on the edge of the painted blue sky. Suddenly, I could see the three silver wires above Blondo Street. Nearly seventy-feet in the air, the wires attached to the wooden pole in a "wishbone" arrangement of angled crossbeams (fig 1). I then recalled the feeling of running and jumping lines dilating and constricting above the station wagon window. Edwards, a young boy who watched spiders fly through the forest and grew up to deliver Early America's most famous fire-and-brimstone sermon, "Sinners in the Hands of an Angry God," had jostled from memory the imaginary video game I played in Nebraska in the 1980s.

NEW GAME. After ruminating on my boyhood play with power lines, I composed an essay connecting Edwards's flying spiders to Ralph Waldo Emerson's "great principle of Undulation in nature" and his vision of the world as "the flux of matter over the wires of thought."[4] To the original link from Edwards to Emerson, I added analysis of two fictional portrayals of undulating lines. In James Joyce's *Portrait of the Artist as a Young Man* (1916) Stephen Dedalus stares out a train window in Ireland and begins to form a prayer. The prayer begins as "a trail of foolish words which he made fit the insistent rhythm of the train." The mix of words and train rhythm is then mapped onto the passing landscape as "silently, at intervals of four seconds,

1. 69kV transmission line, 2018. The three slanted arms attached to the pole form what some utility line workers call a "wishbone" configuration. I began the imaginary "power lines" game at this corner of Blondo and 102nd Street in Omaha, Nebraska. Photo by Eva Hernandez.

the telegraph poles held the galloping notes of the music between punctual bars."[5] Similarly, in Vladimir Nabokov's short story "A Matter of Chance," first published in Russian in 1924, the movement of the wires outside the Berlin-Paris express train reflects a hopeless dining car attendant's mood swings. First, the "even row of telegraph wires could be seen swooping upward," then "a telegraph pole, black against the sunset, flew past, interrupting the smooth assent of the wires." Outside the window the wires, expectedly, "dropped as a flag drops when the wind stops blowing. Then furtively they began rising again."[6] The cadence of the rising and falling lines posted alongside train tracks, I argued, had been reflected in, and possibly an inspiration for, these authors' development of the narrative mode named stream of consciousness. The clean, repetitious undulations in the landscape provided a counterbalance to the narrator's more jumbled, sometimes chaotic thoughts. The perceptions of telegraph lines modeled a somewhat accidental and yet easily identifiable feeling of order. Stephen and the suicidal attendant each desperately sought to latch onto such feelings and the accompanying forms. The loose connections between Edwards and three literary giants—Emerson, Joyce, and Nabokov—marked the beginning of a new game. A series of conceptual jumps and archival sidetracks related to telegraph lines began to dominate my research, writing, and figurative power lines play.

In the years following that Reading Room realization, I began to explore telegraph lines as material and metaphorical conductors of nineteenth-century landscapes. Drawing from a range of relatively static texts and images, I argued that the electric telegraph, and its lines, occupied the razor's edge between science and spirituality, technology and progress. My subjects felt simultaneously unified and polarized: electricity and landscape, power and lines, function and form. With these groupings as my guidelines, I made gentle swoops from the first system (electricity, power, and function) to the other (landscape, lines, and form). I learned that telegraph lines initiated the basic patterns by which electric networks permeated American culture and that their presence in the material landscape also challenged the values that Emerson and others of the American Renaissance attributed to the raw, sudden, electrifying experience of nature. The project showed the diverse iconography of American landscape intersecting with the material and cultural changes wrought by telegraph lines.

Over time I expanded my study beyond the framing of electric landscapes performed in nineteenth-century essays, novels, newspapers, poems, and paintings. On the one side my new research engaged a broader range of "electric" texts, including classic films, utility advertising campaigns, industrial design projects, and made-for-television movies. On the other side it acknowledged that each overhead line represents technological choices and social impacts. Overhead lines, from our high-voltage transmission towers to the nineteenth-century telegraphs, are socially constructed artifacts. To the layperson they look similar, but they can serve distinct cultural functions and accrue separate meanings. For almost two hundred years landscapes lined with overhead wires have acted as fulcrums for the forces of art, culture, technology, and environment.

Sections of the narrative that follow read (and jump) between the lines, so to speak, but my broader agenda is to pick apart and reconstruct how electric lines transmit physical and figurative powers. The interdisciplinary approach is more fully addressed in chapter 1, but my title gestures toward this juncture of the material lines and perceptions of their impact on the landscape. In the common form *power* is an adjective that describes "lines." The lines transmit electric power between points. Although journalists and newspapers sometimes unite the two words as *powerlines*, the intentionally hyphenated compound *power-lined* sets *lined* as the participle and adjective. The *lined* in my title acknowledges the process of lining landscapes and the viewers' experiences of lined space. To focus on ways that places are lined by power is to acknowledge some of the multifarious electric powers, aesthetic powers, and rhetorical powers of electrification. *Power-lined* draws attention to the tensions on either end of the line as well as to the lines' impacts on the visible landscape.

Three general observations ground my study of power-lined landscapes:

1. Overhead telegraph, telephone, and electric power lines form circuits that transmit electricity across significant distances. These circuits have drastically different histories and uses, but by definition each type of electric "line" includes the wires or cables as well as the conductors, support beams, crossarms, insulators, transistors, joints, transformers, poles, guy wires, lightning arresters, towers, bases, platforms, and other tools, materials, and gadgets required to transmit and distribute electric currents. Since the 1840s electricians, engineers, and linemen have built, deciphered, and modified increasingly powerful, broad, and technologically advanced telegraph, telephone, and electric power lines.
2. For the sake of cost and efficiency, the majority of long-distance lines have been strung on upright structures such as poles, pylons, or towers.

3. Viewers will likely recognize the signature catenary shape, the "sag" that allows the cables or wires to shift slightly, but for the majority of Americans overhead lines seem like mere industrial stuff. The lines make ubiquitous, banal impacts on the visual landscape. If the lights stay on, if the network functions, then the material "nets" may be readily dismissed.

These are basic claims. Beginning with the telegraph, modern communication seems to require overhead wires to serve increasingly broader, higher-voltage networks. The electrification of streetlights, homes, and businesses also seems to require overhead wires. Even our modern wireless devices must be recharged—wireless requires wires. In the coming decades new superconductive materials and devices that report and predict electricity flows in real-time should lead to a more resilient and sustainable grid—the future smart grid will probably require overhead wires. Wireless transmission of power is not yet a reality. Wires seem like they are here to stay. Without a clear alternative, human civilization will most likely continue to live alongside, and be dependent upon, overhead wires.

Wires have been strung overhead across the landscape due to technological and financial constraints. Telephone lines or transmission lines buried underground may not be as susceptible to damage from storms, overgrown trees, or vehicular accidents, but when they do fail, it can be more difficult to access the buried lines to make repairs. In addition, undergrounding adds to the significant investment each line requires. In the United States planning, approving, and building a transmission line takes an average of ten years. (Transmission lines operate at higher voltages and usually span longer distances than the distribution lines one sees entering homes and businesses.) Building a 500-kilovolt transmission line costs, on average, $1.9 million per mile.[7] The estimates for running transmission lines underground is regularly six to ten times as much the overhead alternative. In urban areas and places with rocky terrain, an underground line can cost twenty times more than an overhead one.[8] Since the age of the telegraph, when private industry began to build the nation's electric networks, incurring the extra cost of putting lines underground has been seen an unnecessary expense to be added to an already capital-intensive investment.

Many Americans have accepted the overhead lines as a necessary evil. They have learned to ignore or conceal electrically lined environments, pushing wires behind appliances, beneath desks, and behind walls. Hiding our wires reflects a general impulse to forget about electric infrastructure altogether. One reason we ignore the wires that power our lives is the complexity of the physical and social networks that govern the grid. The laws of physics and millions of technologies control the flow of electric currents we use every day. Meanwhile, convoluted organizations such as utility holding companies and regional transmission authorities plan, generate, finance, and regulate the massive currents coursing through these lines. Another reason why we tend to overlook the lines that weave their way across the landscape to the buttons and switches at our fingertips is that electrical engineers, system designers, and landscape architects attempt to limit the lines' environmental exposure and visual salience. They strive to create closed systems, networks that are functional, profitable, and predictable. They also, when possible, attempt to appease collective worries that the wave of visible wires flowing through our streets, parks, and backyards may never ebb. The implicit message sent by overhead lines is "Do not touch and, preferably, do not look." When the lines outside do poke into our attention, we would prefer them buried out of sight and out of mind.

The seemingly universal distaste for overhead lines' aesthetic impact is not a recent development. If transmission, distribution, and telephone lines are grouped with their archetypal predecessors—telegraph lines— then coordinated public resistance to overhead wire blight began as early as the 1870s. When telegraph wires blanketed the streets of New York City, vocal citizens demanded the cityscape be exorcized of so-called wire evil. Almost one hundred and fifty years later, complex electrical systems are still modified to appease collective aesthetic tastes. The U.S. Energy Information Administration reported in 2012 that "nearly all new residential and commercial developments have underground utility infrastructure, often required by law for aesthetic reasons."[9] Laws are passed so we will not have to look at the ugly artifacts that our lifestyle requires. These three observations—large-scale electric networks require lines; lines have often been strung overhead; and since the nineteenth century, many Americans have preferred them to be less visible—become barbed facts when one recognizes that negative perceptions of overhead electric infrastructure are ornery obstacles to large-scale energy reform.

The United States is in desperate need of new and better transmission lines. Transmission lines carry higher voltages, often above 161,000 volts, between generators and substations. Eventually, the higher voltage, which might be considered like the pressure in a water pipe, is stepped down to serve distribution lines and then stepped down again until the electricity arrives at light systems and wall sockets. Most homes are outfitted for pressures of 120 or 240 volts, or about 1 percent of the voltage in a typical transmission line. New transmission lines are crucial to maintaining and improving the nation's aging, inefficient, and vulnerable transmission grid. One valuable intervention to our sweeping infrastructure problems might be called "grid literacy." Several recent titles, especially Gretchen Bakke's *The Grid: The Fraying Wires between America and Our Energy Future* (2016) and Julie Cohn's *The Grid: Biography of an American Technology* (2017), enhance grid literacy and explain the complex relationships between technologies and stakeholders (including engineers, politicians, investors, executives, regulators, environmentalists, and consumers) that helped to electrify the nation. Bakke's narrative focuses on the disruptions between utilities and energy users. Cohn's recounts the history of interconnection efforts and the repeated calls for a national grid that might ship power coast to coast. Both authors recognize the power lines that intersect the landscape, but they also imply that, despite increasing grid literacy, Americans' opposition to overhead power lines continues to impede progress.

Public resistance to power lines can lead to construction delays, numbing litigation, frustrated engineers, and angry citizens. Resistance is not homogeneous; concerns about health impacts, drops in property values, or dangers to habitat and wildlife also fuel concerns.[10] Chapters 4 and 5 address varied impacts and forms of resistance; however, the most widespread, difficult to pinpoint, and seemingly intractable point of opposition to overhead lines is their visual or aesthetic impact. Bruce Wollenberg, a professor of electrical and computer engineering at the University of Minnesota, explains the opposition in blunt terms: "People don't want power lines—period. They don't like the way they look, they don't like a lot of things. It's universal across the country, and I think across the world. People don't want power lines. They don't want more power lines."[11] People want, and need, safe, reliable, inexpensive communication and power, but people do not want to *see* the lines that this privilege requires.

Of course, blanket claims such as "people don't want power lines" or arguments that suggest that the resistance is merely a matter of aesthetic preference can shift too much responsibility onto the viewer. Such claims can lead one to dismiss resistance as ignorance (e.g., "If they understood electricity and business, they would not complain") or all opponents as NIMBY, standing for "Not in My Back Yard." NIMBYism implies that naysayers expect the benefits of a shared resource but will not tolerate any infringement on their personal lives or property. The technical specifications of each line, each landscape, and each public engagement effort are distinct. We might all benefit by increasing our technical and cultural understanding of the technologies and infrastructures that pass through our yards, enter our homes and businesses, and upon which we rely for food, water, energy, transportation, and information. Scrutinizing the immediate systems through which we produce and consume energy can highlight our capacity to create large-scale, positive change.

GAME OVER. My research of power-lined landscapes began as a way to help viewers and communities find new ways to appreciate the lines in their landscape and, potentially, accept the new lines necessary for grid reform. A turning point occurred when I "lost" the power lines game. On June 11, 2012, I stood in Coral Ridge Park, a well-manicured green space in the suburban city of Chino Hills, approximately thirty-five miles east of Los Angeles. Like Henry David Thoreau had done in the 1850s, I touched, peered up, and pondered the existence of a new structure meant to hold electric lines. In *Walden* Thoreau famously criticized telegraphic communication and other "modern improvements" as "improved means to unimproved ends." He also listened, quite literally, to telegraph lines. The strange sounds of this "telegraph harp," he notes in his journal, "always intoxicates me, makes me sane, reverses my view of things. I am pledged to it."[12] That day in Chino Hills I touched the base of a 198-foot tubular steel pole that had been bolted into place but not yet strung by wires. Instead of an eerie whine or high-voltage corona discharge, a hot San Bernardino breeze swept across the sun-drenched valley. I craned my neck and beheld the whole—the pole rose twenty stories, and near the peak, three sets of curved arms stretched to a 60-foot wingspan. Each arm had a stringing block, a wheel waiting to pull cables across the sky. An uninterrupted row of similar steel configurations lined up in the distance and split the patchwork of two-story homes and winding cul-de-sacs (fig. 2). The towers' visual impact reflected a community split over if and how they might fight the line and its owner, Southern California Edison.

The 198-foot towers were part of Southern California Edison's Tehachapi Renewable Transmission Project. The Tehachapi Project would provide a remarkable link in the region's cutting-edge energy infrastructure. At the time it was the nation's largest transmission project devoted primarily to renewable energy. At capacity the 173 miles of new and upgraded transmission lines could transmit 4,500 megawatts, enough wind and solar energy to power approximately three million homes and help to achieve California's ambitious goal of using 33 percent renewable energy by 2020. The environmental benefits of this $2.4 billion power line project seemed unprecedented; however, a 3.5-mile segment of the Tehachapi Project passed through a narrow suburban right-of-way in Chino Hills.

Despite strong opposition from the residents and businesses of Chino Hills, the route and design for the segment through the city was approved in 2009. Construction began the next year. However, in late 2011, after the massive towers had been erected but before conductive cables could be strung and electrified, a court order from the California Public Utilities Commission (CPUC) paused construction. The final tower height of 198 feet meant the line required a separate review by Federal Aviation Administration (FAA). This review would determine where to place orange marker balls and flashing red lights to alert airplanes to the presence of the new, taller lines. The FAA review provided opponents of the Chino Hills segment with a game-changing delay.

2. Steel towers for 500kV transmission line in Chino Hills, California, 2014. The 198-foot towers formed a segment of the Tehachapi Renewable Transmission Project. A court order later forced Southern California Edison to remove the towers and run this section of the 500kV line underground. Photo by Thomas Cordova.

When I first visited Chino Hills, the line was nearly ready to be energized; based on what I had read about it in newspapers before my trip, I thought the line should proceed as planned. It was not perfect, but the environmental mandate must have precedence. Then, as my eyes scanned from tower to tower, I realized that none of the mental acrobatics and alternative views of power lines I had performed in my past had prepared me to accept these intimidating structures. As I spoke with residents and learned more about the circuitous means that utilities can take to achieve quasi-environmental ends, I felt as if the figurative player in my power lines game had jumped . . . and fallen flat on his face. This line may have been built with cutting-edge technologies and the overall intention of reducing carbon emissions, but the planning process did not account for the line's severe impact on the landscape. When one imagined the planned pack of electrified cables beaded with orange marker balls and studded with flashing lights, it seemed clear that this space would be horrifically disfigured by the power line.

The overhead segment in Chino Hills never fully materialized. In July 2013, after a long court battle and a vigorous grassroots campaign, the CPUC reversed its original decision. Five lattice steel towers and eleven tubular steel towers within the residential area of Chino Hills were removed. The conflict in Chino Hills is more fully described in chapter 5, but the feelings generated by those towers made me more sensitive to the massive and often opaque powers that continue to embed and control the energy infrastructure in our environment.

RESTART. I continue to observe, question, protest, write, and play power lines. I promote grid literacy, but I also recognize that other teachers and experts are better equipped to advance scientific and technological understanding of electricity and electrical systems. We need to learn more about the herculean efforts of engineers, line workers, environmentalists, politicians, and policy researchers who maintain current lines and who have devoted their lives to building new, better energy infrastructure. Yet I also believe that using the humanities to contextualize electric infrastructure's wide-ranging impacts can help to garner public support (or at least tolerance) for more egalitarian and ethically responsible infrastructure. Protecting our planet and maintaining our quality of life will likely require new, more powerful, and more advanced overhead power lines. More, and different, lines should appear upon the American landscape. However, I am not a power lines apologist. Some lines have been, and will be, ill conceived, misplaced, and rightfully resisted. Here I attempt not to take sides in favor of exploring the traditions, values, and functions that overhead lines enhance and disrupt as they intersect the American landscape.

A project that began with a boy following a series of metallic menisci outside a station wagon window has swept across the archive and the continent. I have inspected pioneer power lines near Niagara Falls, felt the sublimity of lattice steel towers cliff-hanging over Hoover Dam, heard crackling kilovolts gliding over Minnesota farms, and been appalled by 198-foot "monsters" in Southern California. Now the circuit has closed. After a decade in New York City and four years in Santa Barbara, I moved back to Omaha with my wife and our two daughters to finish this manuscript.

Nebraska is a fitting place to figuratively "beat" my power lines game. In the mid-1840s, when the electric telegraph was new, few Americans understood how it could carry messages so quickly. Many thought the insulated wires were like thin tubes and that the messages could be rolled up and blown to their destinations. An anecdote repeated in national magazines recalled a foolish farmer near Lincoln, Nebraska, who "walked three miles" and stared at the telegraph lines, waiting to "see the man run along the wires with the letter bags."[13] The "man" that this anonymous (and possibly fictitious) farmer expected to see running across the telegraph line seems similar to the imaginary guy with whom I played power lines. Although different lines of thinking inspired our respective imaginations, I'd like to think that viewing overhead electric lines as tightropes or as an undulating running path is part of my Great Plains DNA.

When we moved back to Nebraska, my family and I lived with my father in the split-level house wedged between Blondo Street and Interstate 680 where I was born and raised. The distribution line that brings power to the house links to a transmission line that stems from the Omaha Public Power District Blondo Street substation less than a mile away. That substation, and the broader impact of public perceptions, inspires chapter 6, the last level in my allegorical game.

This afternoon the substation's constant electric drone is being canceled out by the crashing waves of interstate traffic. A 140-foot-tall lattice steel tower is visible across the interstate. Adding for elevation (approximately 1,200 feet), it is one of the tallest structures in the broad, flat plain that stretches toward greater Omaha. I stand on the front porch and look across the stream of flickering cars and semis to the transmission tower, which stands in a thawing cornfield. Like Don Quixote gazing at a windmill on the horizon, I squint at the cellular antennae on the tower's shoulders and imagine they are a knight's armor. The giant's torso twists away from my line of sight, one arm stretches southward, the other to the northwest, both outstretched arms buzzing with tension, challenging me to finish the game.

First, press PLAY.

POWER-LINED

Introduction

In 1832, during a transatlantic voyage, the artist Samuel Finley Breese had a eureka moment. Morse had engaged fellow passengers in a conversation about electricity. Years later he described how the conversation made him realize that "if the presence of electricity can be made visible in any part of the circuit," then intelligence can be "transmitted instantaneously by electricity."[1] That sudden, brilliant *if* initiated the lining of the American landscape.

By the time the packet ship *Sully* docked in New York, Morse was obsessed with electrically transmitting intelligence. He converted his art studio into an electrical workshop, where easels and paintbrushes competed for space with batteries, spools of wire, and paper ribbons that future generations would call "ticker tape." After five years of experiments and tinkering, on September 2, 1837, Morse displayed his prototype, an apparatus built with copper wire and a wooden canvas stretcher. Using a lever, Morse could open and close the circuit, which consisted of a mile of looped wire. Breaks in the circuit were "made visible" as a magnetized iron pen zigged and zagged across strips of paper. A young man in the audience that day, Alfred Vail, became Morse's assistant. Vail made crucial improvements to the machine. He helped transform the lever used to open and close the circuit into the telegraph "key" and the V-shaped markings into the dot and dash cipher referred to as "Morse code," or simply "Morse." Despite additional innovations such as a "relay" to help the signal travel farther distances, Morse's enthusiastic self-promotion in the press, and a success-

1

ful inspection by President Martin Van Buren in 1838, Morse's telegraph seemed more of a curious parlor trick than a practical tool.

Morse's break came in the spring of 1843, when Congress considered a bill to appropriate thirty thousand dollars for Morse to build an experimental telegraph line. During the floor debate Tennessee congressmen Cave Johnson scoffed that if the United States government was to fund a plan as ludicrous as sending messages through an electric wire, then it might as well fund research into mind reading. As one of the first devices to put the mysterious fluid of electricity to use, the telegraph inspired suspicion and wonder. Fortunately, the appropriation passed through House and Senate. Morse confidently predicted "the whole surface of this country" would be "channeled for those nerves to diffuse, with the speed of thought, a knowledge of all that is occurring throughout the land, making, in fact, one neighborhood of the country."[2] Now he prepared to lay the first nerve between Baltimore and Washington DC.

The telegraph keys that sent and received the coded messages represented electricity made visible; the connecting wires, according to Morse's initial plan, would not. Morse wanted to channel his telegraph lines underground to protect them from inclement weather and vandals. Therefore, after securing government monies, Morse hired a team of workers to dig trenches and lay tubes with insulated copper wires. In October 1843 the first contractor's lead tubes corroded. Then the wire's insulation failed. By December, Morse was behind schedule, short of funds, and facing financial lawsuits and a breach of contract. He realized his error, and to gain an excuse for the delay, he asked Ezra Cornell (who would create Western Union and cofound Cornell University) to sabotage his own trenching machinery. Morse then instructed Vail to order "stout spars, of some thirty feet in height, well planted into the ground . . . along the tops of which the circuit might be stretched."[3] Approximately five hundred chestnut poles were placed alongside the Baltimore and Ohio Railroad tracks about sixty-six yards apart. The number 16 copper wire was insulated with a mixture of asphaltum, beeswax, resin, and linseed oil and held onto the poles with glass plates.[4] The overhead forty-mile circuit worked.

On May 24, 1844, Morse sat in the Supreme Court Chambers of the Capitol Building and used his telegraph key to make breaks in a 50-milliampere current (a fraction of the current that typically passes through a wall outlet to charge a smartphone). Vail, sitting in the Monte Clare train station outside Baltimore, which housed an 80-volt battery, recorded the message and sent it back. The ciphered message was a biblical quote, Numbers 23:23—"What Hath God Wrought." With this ambiguous phrase, one that can be read as both a statement and a question, Morse unleashed the so-called electrical age.

The invention of the electromagnetic telegraph and subsequent electric networks has had truly astonishing consequences. Inventors, engineers, and electricians have made invisible electric currents "visible" in telegraph keys, telephones, incandescent bulbs, and increasingly complex circuits, transistors, and microprocessors that keep humankind connected across the cosmos. The success of electric telegraphy wrought a technological revolution that continues today and may be considered as important as any in human history. Morse's success also initiated the proliferation of overhead lines upon the American landscape.

By 1851 more than seventy-five companies were sending electric messages across 21,147 miles of the so-called lightning lines.[5] A decade later, in 1861, at the start of the Civil War between North and South, the first transcontinental telegraph message—"May [this line] be a bond of perpetuity between the states of the Atlantic and those of the Pacific"—passed through thousands of miles of unsettled desert and prairie to reach the East. While lines carried messages and power from sea to shining sea, wiring the continent for telegraph (and later for telephone and electric power) was rarely linear or interconnected. Instead, patchwork networks connected major cities, industrial centers, and urban cores. Some companies and their routes failed because of competition, others due to technological constraints. In 1858 a line that had been dropped to the bottom of the Atlantic Ocean connected North America and Europe for a few months before it permanently failed. In the 1860s a team trying to connect San Francisco to Moscow across the Bering Strait was halted somewhere near the border of Alaska.

Despite the setbacks, by 1880, 291,213 miles of telegraph wire and 34,305 miles of telephone wire connected every contiguous state and spanned the Atlantic. By the turn of the twentieth century the American landscape was lined by 15 million miles of telegraph and telephone lines. Approximately 80 percent were strung from poles, rooftops, or other overhead structures. Another 16,677 miles of wire crossed oceans and rivers.[6] The telephone pushed the telegraph toward obsolescence as the twentieth century advanced, but the spread of electric power systems for street lighting, factories, and household appliances added taller towers and thicker cables to the overhead net. Fast-forward to the twenty first century, and approximately 600,000 miles of transmission lines, 6 million miles of lower-voltage distribution lines, and 1.5 trillion miles of telephone lines connect three hundred million electricity consumers across the continent. Stitched together, these networks form the largest and most interconnected machine in human history.

That Morse's experimental telegraph line could not be buried underground was a curious twist of fate for this aspiring artist and the first professor of painting and sculpture at New York University. Before inventing the telegraph, Morse painted portraits, historical scenes, and American landscapes. His artwork never earned the praise he thought it deserved. One prescient critique of Morse's paintings came from a patron, Phillip Hone. After a large group show in New York City, Hone said Morse's rigid brushstrokes produced "straight lines, which look as if they had been stretched to their utmost tension to form clotheslines."[7] Soon after this show, a bitter Morse quit painting altogether and focused his attention on building an electric device that might transmit intelligence. His invention's success meant that telegraph lines, which looked similar to clotheslines, would stretch across the continent.

Morse's legacy remains tied to his technological contribution. Some contemporaries, however, found humor in the fact that a professor who lectured on the aesthetics of landscape gardening and landscape painting had invented a device that had such severe impacts on the physical landscape. In a cartoon published in 1846, *Yankee Doodle* magazine sarcastically noted the "unity of design" displayed by the telegraph line, which it called Professor Morse's "great national historical work of art" (fig. 3).[8] The cartoon seems to be mocking Morse, who had attempted, and failed, to create great historical works of art with his paintbrush. Then again, the

4

Professor Morse's Great Historical Picture.

YANKEE DOODLE expressed himself much pleased with the unity of design displayed in this great national historical work of art.

3. As a trained portrait painter and first university professor of fine arts, Morse made many attempts to paint grand historical pictures. Ultimately, he failed to achieve the success he desired as a painter, making it easier to abandon his creative ambitions after the successful demonstration of his telegraph in 1844. "Professor Morse's Great Historical Picture," *Yankee Doodle*, October 10, 1846.

telegraph was a great national achievement. In the nineteenth century the image of telegraph lines striding toward the horizon repeatedly symbolized American technology and progress.[9]

Morse's training as an artist helped him to conceive and build his telegraph, but poles connected by dipping wires had not been part of his original design. He did not intend for the overhead line to be viewed as a sculpture or as a subject for landscape painters. Nor was the first overhead line, the one that provided a template for future infrastructure, the first meme in the most pervasive (and likely most horrible) art installation in human history. Or was it?

To consider Morse's inaugural telegraph line as an extension of his artistic practice and therefore accept the familiar form of poles or towers linked by a concatenated wire as a "historical work of art" reveals a deep division between traditional approaches to electric technologies and electric infrastructure. In the popular imagination electric infrastructure is often rendered invisible. Attention is drawn to the machines at either end of the line, those clicking, spinning, and blaring products that put electric currents to work. Morse's inaugural telegraph key and many other historic light bulbs and motors are on permanent display at the National Museum of American History; likewise, an iPod can be seen at the Smithsonian Design Museum. Although many recent studies of electric technologies and culture resist the idea that networked technologies emerge as the result of a single inventor or dominant corporation, they also tend to focus on sites of production and consumption (such as laboratories, telegraph offices, call exchanges, power plants, or street lighting) or histories of specific artifacts (such as batteries, telegraph keys, telegrams, incandescent light bulbs, radios, televisions, cell phones).[10] By comparison, the telegraph, telephone, and power lines, the network's materials and the guts of the grid, have been discarded, relegated to the fringes, or buried out of sight. With a few notable exceptions, such as Nicole Starosielski's *Undersea Network* (2015), electric lines have been pushed to the margins of academic discourse.

Power-Lined brings electric lines to the center. Like a single pole with cables radiating in different directions, this narrative about the webbing of the American landscapes points in at least two disciplinary directions. The history of technology and culture—and more specifically, the history of overhead electric networks—constitutes one direction. The technical development of electric telegraph, telephone, and power lines can be gleaned from accounts of noteworthy projects published in newspapers, history books, and trade journals. These records suggest that competing systems and interests, including aesthetics, influenced the lines' construction and their reception. Owners and promoters used images of the telegraph and power lines to sell visions of their products and systems; opponents often pointed to similar images as signs of blight and degradation. Conductive threads transmitted currents between network nodes; they also reflected the intersecting meanings and values of places lined by electric wires.

The second direction of my narrative considers the lines in relation to the development of the American landscape. Painters, poets, novelists, and philosophers also viewed, interpreted, and appropriated the lines in their environment. The treatment of lines in these more figurative landscapes speaks further to the lines' rhetorical influence. If, as Morse suggested, the lines signaled an attempt to "make one neighborhood of the country," then the presence of lines led some viewers to ask what kind of neighborhoods were lined with wires and what unseen forces may own and control them? Some critics of the telegraph questioned whether the telegraph lines indicated democracy and progress or capitalism's expanding web. Reviewing the history of resistance to overhead telegraph, telephone, and electric power lines contributes to a broader understanding of how Americans have engaged both electricity and landscape.

The coupling of these two narrative threads—electricity on one side and landscape on the other—has been inspired by two distinct histories: Daniel Czitrom's *Media and the American Mind* (1982) and Roderick Nash's *Wilderness and the American Mind* (1967). Czitrom shows that Morse's "lightning lines" penetrated nature, separated communication from transportation, and rewired the ways Americans worked, learned, and constructed beliefs. Days after Morse's inauguration of the first telegraph line, the *New York Herald* said the device "has originated in the mind an entirely new class of ideas, a new species of consciousness."[11] Each successive widespread electrical invention—telephones, light bulbs, home appliances, televisions, and computers—has seemed to spark a "new" consciousness along with new orientations of space. Czitrom adds, "New media [telephone, radio, television] reshape our perceptions of the past and the contours of knowledge itself."[12] Media formats the messages that are sent through the wires. Whether the messages transform into dots and dashes or ones and zeros, media codes and decodes electrical impulses. Media technologies, like infrastructure, also shape human relationships to knowledge and consciousness. Devices connected by wires are not merely, as Morse suggested, the conduits of "intelligence made visible"; they signify intelligence and literally shape how it is transmitted between senders and receivers. From telegraphs to killer apps, electricity influences what and how we know about our ancestors, our universe, and ourselves.

Electricity's force in American history has been mapped far beyond its scientific underpinnings and technological applications. Czitrom, Thomas P. Hughes, David Nye, Linda Simon, David Hochfelder, Jill Jonnes, and Daniel French have each displayed electric technology's revolutionary impact on American finance, commerce, politics, war, journalism, and entertainment.[13] Literary scholars have also analyzed electricity's dynamic relationship to the arts and sciences. The power and flexibility of electricity-as-metaphor gave rise to what James Delbourgo calls the "political electricity" of the Enlightenment, what Paul Gilmore calls the "aesthetic electricity" of Romanticism, and the "net-works" that Laura Otis argues formed feedback loops between novelists and biologists rethinking communication and the nervous system in the age of Charles Darwin. Sam Halliday suggests electricity in the nineteenth century was something "to think about and with." Thomas Edison and his team used a range of genres, rhetorical devices, and representational systems to invent the allure of artificial light. Twentieth-century novelists including Charlotte Perkins Gilman, Jack London, and Ralph Ellison seized the "polysemous" emblem of electricity because "it already had a rich aesthetic legacy and because its new industrial applications correlated this energy with interconnection and action at a distance."[14] Clearly, electricity interweaves history, language, and culture; it charges how we think, act, and feel. Previous examinations of this multimodal and tenebrous force, however, have not fully accounted for how electric infrastructure and the sight of lines in the landscape influence electric ideas and texts. I revisit some of the authors and texts treated in previous studies to show wires and lines as actors and artifacts.

Only a handful of scholarly articles have analyzed transmission lines as cultural and aesthetic artifacts. In a short piece written for the National Park Service, Leah Glaser argues that transmission towers are more than "a blemish on the landscape"; they are, she notes, also "valuable cultural resources with a crucial story about the impact of long-distance power."[15] In his excellent history of transmission tower designs Eugene Levy reviews the power industry's attempts to reduce public hostility by building modernistic, "aesthetically pleasing" structures. Levy concludes, "Rather than being defined as symbols of industrial progress or as monumental structures that are an integral part of the contemporary land-scape, for most Americans transmission structures remain merely another example of the trashing of the environment."[16] My narrative takes a similar approach, valuing the cultural and symbolic power of the power line expanding the scope to include distribution, telephone, and telegraph lines. Each of these similar-looking types of overhead lines transmit electric currents between points in a circuit; they also transmit subtle and sometimes conflicting messages into the landscape.

If electricity and electric technologies have been one of the most potent physical and rhetorical forces in American culture, constructs of landscape have provided some of the most powerful frames. *Landscape* comes from the Dutch *landschap* and Old English *landship* and combines a familiar noun, *land*, with a variation of the verbs *scap* or *scheppan*, which generally meant "to shape, form, or create." As the term's meaning has evolved, standing at a particular vista, crossing a field with a plow, putting paintbrush to canvas, even looking through a viewfinder, can "create land" into a landscape. To name something "landscape" implies an exercise of certain beliefs and values about space, aesthetics, and environment. As W. J. T. Mitchell has noted, *landscape* is a noun and a verb, "both a represented and presented space, both a signifier and a signified, both a frame and what the frame contains, both a real place and its simulacrum, both a package and the commodity inside the package." This duality and a series of especially flexible and sometimes contradictory frames have created what Mitchell argues is a compelling "instrument of cultural power"—the *American* landscape.[17]

Nash's study of wilderness unearths some of the specific myths and realities of the American landscape. In the popular imagination wilderness is the *absence* of civilization, but repeated encounters with (and projections of) seemingly unsettled lands fostered the rugged individuality and self-reliance that came to define American identity and culture. To retain its wilderness quality, humans cannot stay, but when they leave, they carry qualities of "wildness," or "wilderness," away.[18] A similar influence can be seen in the values attached to, and drawn from, agrarian and pastoral landscapes. According to Leo Marx, Thomas Jefferson believed that the "physical attributes of the land" were less important than its "metaphoric powers" and that "what finally matters most is its function as a landscape—an image in the mind that represents aesthetic, moral, political, and even religious values."[19] Since the early nineteenth century detailing and revising images of American wilderness, frontier, and family farm have been paramount to American conservation and environmentalism. Forester and philosopher Aldo Leopold observed that developing recreational spaces "is not a job of building roads into lovely country, but of building receptivity into the still unlovely human mind."[20] For Leopold building receptivity begins with patience and respect for landscapes and the diffuse and delicate ecosystems they contain. Each era develops new means of understanding and appreciating the American landscape's particular power. Of course, the process of exploring, defining, and revising the meanings and values of a landscape brings with it all kinds of hopes, prejudices, and errors.

However flawed or biased, framings of the American landscape, like our electric technologies, position collective perceptions of space and place. John Conron maintains that the United States has repeatedly renewed this exceptional relationship with landscape. "In no culture has the spatial construct of landscape been more indispensable," Conron argues, "for we seem to see ourselves as a people living in space more than in time, in an environment more than in history."[21] Our history and our worldviews are embedded with spatial reckonings of the immense and unsurpassable, the endless and tame, the awe-inspiring and the mundane. Charles Olson opens *Call Me Ishmael*, another monumental work of American studies, with the statement: "I take SPACE to be the central fact to man born in America, from Folsom cave to now. I spell it large because it comes large here. Large, and without mercy."[22] This large, open, jubilant space, in Olson and elsewhere, is transformed into place by the physical, psychological, and aesthetic process that *creates* land, that is, via landscape.

The overhead electric wires and cables that have ranged, as Morse predicted, "the whole surface of this country" represent a powerful, albeit troubling synthesis of electricity and landscape. The lines that carry electricity through American landscapes seem restrictive, redundant, and prosaic. Nevertheless, these lines represent points where electricity is made *visible*. Closer readings of the American landscapes through which electric lines pass can help to clarify the meanings sent by lines and wires. Conversely, the siting and reception of electric lines reflects how particular groups and communities use and value landscapes.

The challenge of balancing the forces and frames wrought by overhead electric lines may be best exemplified by the opening sentences of Thomas Hughes's groundbreaking history of technology, *Networks of Power* (1983). "Of the great construction projects of the last century," Hughes begins, "none has been more impressive in its technical, economic, and scientific aspects, none has been more influential in its social effects, and none has engaged more thoroughly our constructive instincts and capabilities than the electric power system."[23] The superlatives of the first sentence—"none has been more" and "none has engaged more"—seem appropriate. Electrification was selected as the "greatest engineering project" of the twentieth century.[24] The true scale of the cooperation, ingenuity, and insight required to conceive, construct, and revise the massive electrical infrastructure spanning the earth's surface and pulsing with the lifeblood of modern commerce seems to rival even the wonders of the internet and our explorations of the solar system.

Hughes's summary of electrification's effects is certainly optimistic. One could counter his superlatives by arguing that of all the human-made systems built in the last century, none has had such damaging environmental impacts (e.g., air pollution, light pollution, and sprawl), and none made us so hopelessly dependent on infrastructure about which most of us know so little. Electrification's equivocal powers are made clearer by the second sentence of Hughes's tome, as he moves from a series of superlatives to a totalizing claim: "A great network of power lines which will *forever* order the way in which we live is now superimposed on the industrial world."[25] *Forever* is a curious word choice here, particularly for Hughes, a scholar who helped establish the social construction of technology (scot) theory and the idea of "technological momentum."

The scot approach to technological systems, introduced by Wiebe Bjiker and Trevor Pinch in the 1980s, explains the emergence of technological artifacts in terms of social processes such as "variation and selection," rather than the sheer genius of an inventor or an inherent usefulness of a technology. SCOT works to understand the alternative models and failures that belie successful technologies and surrounding systems. It calls attention to social groups that engaged with the technology and the "interpretative flexibility" that arises from contemporaneous debates and framings of artifacts such as bicycles or light bulbs. Hughes applied SCOT theory to the development of large electrical systems to show how the success of individual artifacts, such as the alternating current motor or electrical outlet, was dependent upon and connected with broader organizational, economic, and political circumstances. Hughes explains, "Persons who build electric light and power systems invent and develop not only generators and transmission lines but also organizational forms as electric manufacturing and utility holding companies."[26] Heterogeneous teams of engineers, electricians, financiers, advertisers, and managers invented and developed pieces of power grids with the intention of imposing a particular kind of order on the world and stabilizing their own lucrative business practices. Of course, the social (and environmental) effects of such systems are not permanently closed, and the interpretative flexibility encouraged by SCOT theories suggests that even seemingly antiquated technologies, such as power lines, can adopt new functions and be laced with new meanings.

The adaptability of power networks also explains Hughes's idea of technological momentum. Hedged between fixed determinism and social constructivism, technological momentum suggests that technologies and their systems are open to change. A prosperous technology begins with a series of technical, scientific, and social interventions and adjustments. Successful decisions help the technological artifact and the system surrounding it to proliferate. As telegraph networks or power grids spread, it becomes more difficult, but not impossible, for subsequent generations to change or replace the system or for the system to perform new functions. Early in the twentieth century electric power shaped society, and in turn, society shaped electric power. However, the construction materials, engineering practices, codes, laws, regulations, corporate structures, and cultural conventions that helped electric power systems expand and multiply increased the technological momentum of a specific kind of grid. In North America, for example, the fact that our grid operates at 60 hertz is one aspect of strong momentum. It would be difficult, but not impossible, to create a different kind of grid that transmitted electric currents at a different frequency just as it would be difficult, but not impossible, to remove all overhead wires and bury them underground. Therefore, the more rigid grid we have inherited is, as Hughes explains, "less shaped by and more a shaper of its environment."[27]

Overhead power lines may order the way humans live "forever." Overhead electric lines (some of which still look like they did in Morse's day) could be more permanent than our homes, railroads, highways, and internet protocols. What does seem clear is that the power systems our lives require will not be easily rewired. In 1934 theorist Lewis Mumford noted, "Wires carrying high tension alternating currents can cut across mountains which no road or vehicle can pass over."[28] These mountain-leaping lines helped to establish the "utility," an entity that Mumford warned would, once established, be entrenched with political and sociotechnical power. Almost ninety years later, the utility, and the grid it owns and controls, has been reluctant to change. As Bakke suggests, our grid today is "a technological monument to recalcitrance."[29]

Power-lined landscapes will not vanish. Electric infrastructure will not readily disappear. Instead, like decrepit buildings, unexploded bombs, or abandoned mines and quarries, wires and poles will likely occupy large swaths of the earth's surface for decades and possible centuries after they become obsolete. The potential success of wireless transmission of power could make the word *wireless* seem superfluous (just as the notion of a wireless telephone or landline has nearly lost all currency). Such a technological revolution could turn power lines into historic artifacts. Certain models may be considered as culturally significant as the wagon ruts across Nebraska or the Roman aqueduct in Segovia, Spain. The obsolete power line will provide a visible example of the pathways, grooves, and channels built by our ancestors. It will remind future generations about the ways a former people lived and worked and viewed the landscape.

How strong is the momentum of our overhead lines? Has our society been permanently power-lined? If our metallic lines and the steel poles and towers do survive thousands of years longer than our civilization, how might be they be interpreted by historians or archaeologists thousands of years in the future? Those investigators, who might represent a distinct species, would likely discover more advanced machines and systems, but how might they account for the fact that metallic threads touched billions of inhabited spaces and extended across deserts, oceans, mountains? What might they imagine was our relationship with the wires with which we webbed our planet?

To better understand the order that power lines impose on the present and how new lines may determine our energy future, it is necessary to track them into the past. The next two chapters of *Power-Lined* follow electricity, landscape, and the wires uniting them as they converge and diverge from the period just before the spread of telegraph networks to the rise of telephone and power networks that eventually made the telegraph lines obsolete. Chapter 1 focuses on the electric aesthetic cultivated by Morse and the other artists and writers of the American Renaissance. Chapter 2 examines "frontier" lines developed in historiography by Frederick Jackson Turner and at Niagara Falls by Nikola Tesla. The state of New York provides two competing visions of nature harnessed by power lines and inundated by a wire forest. Chapter 3 shifts to California to examine telegraph lines in two of the first westerns made in Hollywood, the public reactions to long-distance transmission of power from the Sierra Nevada, and the pressures that led to the first, and last, attempt to design and build aesthetically pleasing power lines. Chapter 4 recounts the history of rural and suburban perceptions as told in Rural Electric Administration materials, popular films, and book-length exposés. It concludes with an analysis of the recent struggle between Southern California Edison and Chino Hills. Finally, the conclusion reviews the current state of the power-lined landscape and calls for a power lines poetics to help balance grid literacy with the other interpretations of lines in the visible environment.

Throughout, my narrative recounts a range of reactions to and representations of overhead lines as they appeared in the American history, prose, poetry, fiction, film, paintings, industrial design, and various elements of popular culture (e.g., editorials, advertisements for electric power, and made-for-television movies). I have only scratched the surface of the scientific and technological developments of telegraph, telephone, or power lines. This narrative is not geared specifically towards line workers, engineers, or utility managers, though of course I welcome them as readers and potential advisors who might help me learn more about their fascinating fields. Instead, my focus rises and falls toward the cultural, rhetorical, aesthetic powers of overhead lines. I believe that by tuning our minds to the physical and conceptual threads that have occupied and defined our landscapes, we might calibrate the seemingly disparate frequencies of large-scale energy systems and environmentalism.

1

Wires in the Garden, 1844–1882

Samuel Morse completed an experimental, 40-mile telegraph line in 1844. By 1853, eighty-nine long-distance connections covered 23,261 miles. In 1866 Western Union made a lasting transatlantic connection, introduced the first telegraph stock ticker, and took control of a vast 76,000-mile transcontinental network.[1] During these unstable decades of development, to watch workers erect wooden posts and pull copper or iron wire was to watch a place be figuratively tied to the United States. To travel through the wilderness and suddenly come upon a series of wired poles in an otherwise vacant landscape was to see the direction where American settlements had been or would soon follow.

Scholarly discourse of technology and American landscape has primarily overlooked telegraph infrastructure. For example, Leo Marx's seminal study *The Machine in the Garden* includes a single passage specific to telegraph lines. In 1850 an English visitor to the United States expressed surprise that lands recently occupied by "wild beasts, and still wilder Indians" had been "traversed in perfect security by these frail wires." The comment reinforces racist divides between "wild" nonwhites and a widespread apparatus for enforcing "perfect security." In addition, Marx suggests, the anecdote implies that in America "progress is a kind of explosion."[2] Such a reading conforms to Marx's overall argument that between the late eighteenth century and early twentieth century, the seemingly sudden appearance of new, innovative machines framed the American pastoral and restructured the "landscape of the psyche." Marx reads this restructuring through the treatments of the mill, the steamboat, the factory, the combustion engine, and especially the railroad by writers, including Thomas Jefferson, Ralph Waldo Emerson, Nathaniel Hawthorne, and Mark Twain.[3]

Closer attention to the creative renderings of telegraph lines can enhance our understanding of nineteenth-century technology and landscape, but these representations do not exactly support Marx's thesis. For example, Marx quotes from Walt Whitman's "Passage to India" (1869)—"In the Old World, the east, the Suez canal / The New by its mighty railroad spann'd, / The seas inlaid with eloquent gentle wires"—to show the "buoyant power" evoked by "the machine's [i.e., the railroad's] motion across the landscape."[4] Whitman's vision of the expressive wires on the seafloors is less buoyant but no less critical in the material links between old world and new.

Lewis Mumford helps to explain the missing thread in Marx's argument. In the early twentieth century people overlooked the power utility and infrastructure because, he said, "attention is directed more easily to noisier and more active parts of the environment."[5] The same was, and is, true of telegraph lines. The static, gentle wire that connected communities and traversed land and sea was not the locomotive beast that chugged coal, shrieked its whistle, and spewed smoke through the garden; it was the corresponding vine that crept through it.

During the first decades of telegraph development, the telegraph line, or simply "the wire," entered a synecdochic relationship with the practice of widespread, seemingly instantaneous communication. Meanwhile, telegraph lines sent conflicting messages through the landscape. Many Americans embraced the telegraph line as a wondrous miracle sent to "annihilate space and time," and in such readings the line signaled industrial and cultural advancement. Others resisted telegraphy, worrying that the iconic, visible web would tangle everything and everyone within its reach. A few individuals registered the infrastructure's subtler effects. Wind playing over the wires, or what is now called the "Aeolian effect," produced a discordant whistling sound that Thoreau said was "fairer news than the journals ever print."[6] As telegraph lines stretched in Western Territories, they connected the farthest extremities of Anglo settlement to the rest of the United States. These nerves, homesteaders and armies learned, could be touched and also broken.

American Landscape, Charged without Wires

Samuel F. B. Morse personifies the crossovers and tensions between electrified landscapes and electric technologies in the mid-nineteenth century. He is most often remembered for his role in the technological revolution. In 1845, when Morse was fifty-three, a law clerk wrote a poem declaring Morse as the one who "yoke[d] the lightning to his rapid car" and would have his name etched "on the same tablet with our Franklin."[7] During the next decades the telegraph lines became colloquially referred to as "Morse lines."[8] In 1871, a year before Morse's death, the poet William Cullen Bryant declared, "Every telegraph wire strung from post to post, as it hums in the wind, murmurs his eulogy."[9] However, in his late thirties, long before Morse lines made their vague, meandering mark on the visible terrain, the painter and professor worked fervently to make his mark as an artist and, as part of these efforts, helped give rise to the genre referred to as "American landscape." For the rest of the nineteenth century painters and poets celebrated both wild and pastoral environments as repositories of organic, electric forces. These celebrations preceded, and then paralleled, the development of telegraph networks.

The genre of American landscape, the first significant school in American painting, might be traced back to 1826, when Morse, Thomas Cole, and Asher Durand founded the National Academy of Design in New York City. Morse was elected as president of the new artist organization, and, in March and April 1826 he delivered a series of lectures at Columbia College titled "On the Affinity of Painting to the Other Fine Arts." Morse defined *landscaping* as the process of "hiding defects by interposing beauties; of correcting the errors of Nature by changing her appearance."[10] For Morse the main objective for the landscape gardener, architect, or painter was to "select from Nature all that is agreeable, and reject or change all that is disagreeable."[11] This aesthetic approach is displayed in Morse's three major landscape paintings: *The View from Apple Hill* (1828–29), *Allegorical Landscape of New York University* (1832–33) (fig. 4), and *Niagara Falls from Table Rock* (1835).

Morse's belief that the artist should reject or correct nature's "errors" ran contrary to the emerging trend in American painting, poetry, and fiction.

4. Morse's idealization of the university shares some stylistic qualities with his contemporary Thomas Cole's *Course of Empire* (1833–36). Samuel F. B. Morse, *Landscape Composition: Helicon and Aganippe (Allegorical Landscape of New York University)*, 1836. Oil on canvas, 22½ x 36¼ in. New York Historical Society.

For instance, Morse's *Allegorical Landscape* shows a formalist selection of nature's parts that follows his belief that the artist corrected errors and created an ethical message: nature, virtuous and tame. Cole, Durand, and other now-famous American artists and writers such as Emerson, Thoreau, and Walt Whitman embraced the bare, unruly, and abundant wilderness: nature, red in tooth and claw.

In 1831, after traveling abroad with Thomas Cole to paint pastoral scenes in Italy and France, Morse stopped in Paris and stayed with James Fenimore Cooper and his family to finish his most ambitious painting, *Gallery of the Louvre* (1831–33). Cooper had arrived in the French capital in 1826, on the heels of his most successful novel, *Last of the Mohicans* (1826). Cooper stayed overseas to write three more novels in the five-part Leatherstocking Tales, each featuring the trapper-turned-frontiersman Natty Bumppo. With these novels Cooper successfully packaged American ruggedness for European audiences; with *Gallery of the Louvre*, a massive painting re-creating thirty-eight masterpieces in a single canvas, Morse wanted to bring European taste to the United States.

As Morse painted in Paris, Durand and Bryant organized a picture book in New York City that they titled *The American Landscape* (1830). Durand and Bryant claimed their collaboration was the first to offer Americans "accurate views" of places such as the Delaware Water Gap, Catskill Mountains, and "Winnipiseogee Lake" (Winnipesaukee Lake). The collection of faithful drawings, historical allegory, and poetic illustrations crucially lacked those "tamings and softenings of cultivation" that "change the general face of the landscape" and "break up the unity of its effect." The absence of civilization would remain crucial to the idealized American landscape. Meanwhile, formal landscape aesthetics may have been developed in Europe and epitomized by some of the paintings Morse copied in the Louvre, but the "perception of [nature's] charms is not less quick and vivid among our countrymen." For the American artist a perceptive eye and adventurous spirit can compensate for any lack of aesthetic training or long-standing tradition. Therefore, sharing such charming, rugged landscapes with fellow Americans, Bryant explained, is a means to "promote the success of an experiment hitherto untried, and perhaps hazardous."[12] In other words, the experiment initiated by breaking free from European control and forming the United States gave rise to an experimental view of the North American continent. If *landscape* was a verb, then capturing and promoting the raw, wild, untamed American landscape was an act of patriotism.

Notwithstanding Morse's own corrosive politics (he was a paranoid, anti-Catholic, anti-immigrant politician and a rabid conspiracy theorist), the professor had an ardent desire for public appreciation—and funding. Morse spoke about the difficult life of the starving artist on the first anniversary of the National Academy's formation in 1827. As its president and the self-appointed champion of "Public Taste" in the United States, Morse, in his speech, stressed the artists' collective need to raise the general level of appreciation for the arts and to create a market for American paintings and sculptures. If artists could teach the general public about the true value of art, then patrons may begin to pay for it. Educating the unruly masses about the finer qualities of art would be difficult. In fact, Morse painted a rather dim picture of the art market and then asked his fellow artists: "Why do I speak to you of difficulties? For they are the glory of genius, without which its energy and its brilliancy would pass unnoticed away, like the electric fluid which flows unobserved along the smooth conductor, but when its course is thwarted, then, and only then, it bursts forth with its splendor, and astonishes by its power."[13]

Some of the more talented and resilient artists listening to Morse's National Academy speech, such as Cole and Durand, were able to tap into the forces that "flowed unobserved" through nature and help them "burst forth" with beauty and splendor on the canvas. The electric qualities of plein air painting appear in Cole's "Essay on American Scenery" of 1836. Far from the city, Cole feels "the quickening spirit" of nature. He says that during one moment of sublimity, "such as I have rarely felt," he feels the rocks, wood, and water "brooded the spirit of repose," and "the silent energy of nature stirred the soul to its inmost depths."[14] Pure, unfettered energy captivates the artist, who then transmits the electric experience onto the canvas.

Morse curated and promoted artists who visualized untamed nature's electrifying effects; Emerson verbally conducted electric landscapes. In "Literary Ethics" (1838), for example, Emerson said the American artist will someday, "like the charged cloud, overflow with terrible beauty, and emit lightnings on all beholders."[15] Morse used a controlled, technological image (smooth conductor), and Emerson offered an effervescent, natural image (charged cloud), but both implied that when American painters and poets do achieve greatness, their work will be like sparks erupting from a circuit or bolts tearing across the sky.

Although Emerson's fascination with electricity and electric thinking has been well documented, these ideas are also germane to his treatment of landscapes.[16] Solitary, titillating sensations suffuse Emerson's experiences with the outdoors, especially his seminal text, "Nature." *Landscape* appears four times in that essay, more often than the words *God* (3), *poet* (3), *mind* (3), or *beauty* (2). The only words Emerson repeats more often are *nature* (13) and *man* (12). The various inflections of Emerson's "landscapes" merge in his most electric, transcendental experience. Walking across the bare common, he declares: "I become a transparent eye-ball; I am nothing; I see all; the currents of the Universal Being circulate through me; I am part or particle of God."[17] These invisible, circulating currents and particles seem to underlie Emerson's earlier reference to the "charming landscape" made of lands owned by farmers—"none of [whom] owns the landscape." The emotional transfer between self and environment lingers in the "tranquil landscape" and the "contempt of the landscape felt by him who has just lost by death a dear friend." Landscape for him is an attractive, unstable, invisible force that charms a sensitive perceiver. It is also like a light beam that the individual can project onto nature. For Emerson the most powerful landscapes, like electricity, are fusions of matter and spirit, of human and nature.

In 1844, the same year Morse sent his inaugural message, Emerson declared that "the Poet" had a "power transcending all limit and privacy" and was thereby a "conductor of the whole river of electricity."[18] Inflections of Emerson's poet-as-conductor and popular beliefs in animal magnetism—the theory that some individuals could control electric forces and thereby hypnotize or sexually attract others—collide in Whitman's famous "I Sing the Body Electric" from *Leaves of Grass* (1855). In Whitman's poem the electric rhetoric is conveyed through evocative, erotic images: "Ebb stung by the flow and flow stung by the ebb, love-flesh swelling and deliciously aching, / Limitless limpid jets of love hot and enormous, quivering jelly of love, white-blow and delirious juice." The electric juice flows body to body—no wires required. Significant attention has also been called to Whitman's idea of the "body electric," but acknowledging how Whitman's language of electricity (*conductor, current, shock*) entwines with his depictions of landscape (*field, scene, vista*) enhances such readings. In "Song of Myself," for example, Whitman feels the currents of the masses coursing through him and boasts: "I have instant conductors all over me whether I pass or stop / They seize every object and lead it harmlessly through me."[19]

Approximately forty lines later, Whitman's electric, licentious sensations dissipate. As the energy that flowed in seems to ebb away, Whitman "stand[s] by the curb" and bears witness to "prolific and vital, / Landscapes projected masculine, full-sized and golden." Like Thoreau, who states in

Walden, or Life in the Woods (1854), "Wherever I sat, there I might live, and the landscape radiated from me accordingly,"[20] Whitman suggests landscape is more of a quality than an object, something that can be generated from within and radiated outward. To create this kind of esoteric, electric landscape, the poet draws the current pulsing through nature into her body and projects it back onto the environment.

Morse's contributions to American art and culture deserve recognition, especially his portraits, his efforts to organize and educate aspiring artists, and his introduction of the daguerreotype to the United States. However, whereas his fellow painters and poets such as Emerson and Whitman embraced the wild, vulnerable, and electrifying, Morse continued to push for the refined, elevated, and conservative. Eventually, Morse abandoned his artistic pursuits. He earned wealth and fame from his telegraph, but he remained bitter about his artistic failures. He wrote in a letter to Cooper years later, "Painting has been a smiling mistress to many, but she has been a cruel jilt to me."[21] Morse's art was unable to channel and transmit invisible, all-pervasive electric forces in the sublime and untamed; instead, he invented a telegraph that lined American landscapes with wires.

Electric Binds

In approximately two decades telegraphy changed American politics, journalism, finance, commerce, and war. These effects corresponded to the lines that Americans envisioned and saw stretched through the material landscape.

In June 1844, weeks after the inauguration of the nation's first telegraph line and before plans for further lines had materialized, the *New York Herald* predicted that telegraph would "bind together with electric forces the whole Republic" and thereby "do more to guard against disunion" than any patriotic government. National cohesion and the flow of people, goods, and information would be facilitated with "this great, subtle, wonder-working element of electricity everywhere ready to do our bidding." The telegraph would not work alone, but it would provide "the soul of the vast framework" of roads, rivers, ports, and railroads.[22] Such comments follow Morse's claim that the telegraph would make "one neighborhood of the country" and also foreshadow ensuing assertions that the telegraph represented the "manifest destiny" of Anglocentric advance.

The notorious term emerged in 1845. As telegraph lines reached from Baltimore to Philadelphia and then to New York, John O'Sullivan's *Democratic Review* argued that the annexation of Texas, and eventually all lands to the Pacific, was the predetermined and natural result of "the manifest destiny to overspread the continent allotted by Providence." The expansionist ideology underlying manifest destiny had dictated development patterns for centuries. God had granted the American people wondrous technologies and a moral high ground; they were destined to use their talents and occupy the continent. Not only was this national growth a spiritual duty; it was efficient. O'Sullivan argued that other neighboring nations, particularly Mexico, could not build such a "vast skeleton framework of railroads, and an infinitely ramified nervous system of magnetic telegraphs."[23] Even those indigenous peoples displaced by the spread of this utopian destiny stood to benefit from the technological, scientific, and cultural advancements that Americans brought with them. Colonization was veiled as altruism.

Manifest destiny explained, and often excused, the momentum guiding the settlers of the West. The telegraph was particularly effective at helping Americans visualize this push. Seven years after O'Sullivan coined the phrase *manifest destiny*, for example, the *American Telegraph Magazine* suggested it was "manifest destiny" that was "leading the 'lightning' abroad over this capacious continent of ours." The lines acted like a magnet, pulling settlers and commerce westward and then keeping the nation bound together, "not merely by political institutions, but by a Telegraph and Lightning–like affinity of intelligence and sympathy."[24] Trade, government treaties, and armed forces helped protect American interests, but the telegraph lines kept the nation in touch, physically and emotionally.

In addition to supporting expansionist ideologies, in 1846 the *New York Evangelist* predicted that in a few years "a vast revolution will have been affected in the newspaper business through the medium of the magnetic telegraph."[25] Journalists and editors quickly learned that securing the latest telegraph dispatches from the wild frontier or the violent frontlines helped sell newspapers. The telegraph thus influenced the gathering and distribution of news and allowed for regional, national, and even global issues and events to shape local policy debates.[26] During the violent conflict that followed the annexation of Texas—the U.S.-Mexican War (1846–48)—New York newspapers pooled resources to build telegraph lines and hire steamships to carry the latest reports from the frontlines. The collaboration led to the formation of the Associated Press. Again, the manifest destiny facilitated by the construction of telegraph lines across the American landscape enacted a kind of self-fulfilling prophecy. The wires helped initiate and then monitor expansion. As the networks stretched westward, often romanticized reports of "progress" flowed East.

In addition to distributing the latest news, the lines linked rural sites of production and urban sites of consumption. At the same time that the Treaty of Guadalupe Hidalgo ended the U.S. war with Mexico, in 1848, telegraph lines linked New York to Washington DC and on to New Orleans via Richmond, Charleston, and Mobile. Long-distance messages often required relays, and messages could take hours to reach their destination. Direct connections between major cities meant that intrastate conversations might only lag by a matter of minutes or seconds. By the midpoint of the century, however, telegraph lines bordered the Great Lakes—Erie, Cleveland, Toledo, Detroit, and Chicago—and went as far as St. Louis.[27] Farmers could check the price of grain or cotton in a distant market and then decide if they should harvest their crops or hold back in hopes of getting a better price. Goods ready to be shipped could be "sold" before they arrived. Bankers could check a customer's credit from another city or another state.

Some communities were proud to see the lines entering their town, as they indicated a certain social standing. As the telegraph reached from Chicago to Racine and Milwaukee in 1848, excited farmers who lived along the route provided the workers with poles free of charge.[28] In 1848 an Ohio newspaper said the telegraph line from Columbus to Portsmouth was "better constructed than any other in the country, excepting the Townsend telegraph from Baltimore to Wheeling, which is superior to ours in one particular only, that is, the greater size of the posts."[29] For many communities the size and design did not matter as much as the fact that the lines delivered the latest in politics, culture, and commerce well in advance of the mail carriers.

Electric information and electric lines were not always welcomed. Menahem Blondheim suggests that the ability to send messages from distant, unverified sources opened up the "potential prostitution of the [telegraph] invention."[30] False rumors spread, especially in the initial years when newspapers scrambled to get a scoop. In 1845 a dispute in Boston led a New York newspaper to scoff that "fanaticism" and "some petty spite" had caused a group of newspaper editors to call for "the magnetic telegraph poles in that city cut down."[31] Public opinion also turned against the lines when reports surfaced of stock jobbing and manipulation of commodities prices. In 1846 a Charleston newspaper reasoned, "The law may make it a penitentiary offence to break down the wires, but in the present state of public opinion, no jury could be found to convict any one of the offence." It advised readers, "The sooner the posts are taken down the better." In New Orleans the general public had a "most fervent wish that the telegraph may never approach us any nearer than it is at present."[32] In 1850 a suspicious outage was linked to an unknown group of cotton speculators. As such practices primarily did injury to the "planters and raisers," the *Mississippi Creole* said cotton farmers were justified in "tearing down the telegraph poles."[33]

The poles' owners fought to protect their interests. In 1852 Moses Knight and an accomplice were found guilty of cutting telegraph wires in South Carolina. The contractor sent to fix the wires reported back to his superiors that Knight had been sentenced to "thirty-nine lashes on the bare back" and promised thirty-nine more lashes if he ever returned to Marlboro County. Although it is unclear what happened to Knight's coconspirator, the contractor wrote that he believed that this man, and anyone else interfering with the lines, should be hanged.[34]

General suspicions and isolated incidents of sabotage did less to slow development than lack of financial backing, shoddy construction work, and inclement weather. In theory the lightning lines integrated an expanding empire; in practice the "frail wires" erected in the initial bonanza years represented what Robert Luther Thompson describes as "methodless enthusiasm." By 1850 twenty telegraph companies existed, with half of them operating within the state of Ohio.[35] New companies, sometimes with competing telegraph systems or redundant routes, built haphazard lines between places where they felt demand could rise or where they felt they might undercut a competitor's rates. Jokes circulated that in rural areas those trying to get a foothold in a particular market strung shoddy wires on "beanpoles and cornstalks."[36] Batteries routinely failed, circuits shorted, and strong weather (as it still does) brought down poles. Lightning storms also wreaked havoc on the lines and supposedly supernatural forces caused telegraph keys to tap uncontrollably. Despite suspicions and inefficiency, the telegraph lines spread, and by 1851, 21,147 miles of telegraph wire connected approximately 500 cities and towns across the United States and Canada.[37] Telegraph poles studded the Eastern Seaboard, and the growing wire network advanced west of the Mississippi River.

Rather than automatically binding the nation together, the spread of telegraph lines in the 1850s may have increased regionalism and stoked suspicions. In 1861, after a series of secessions and the first shots at Fort Sumter, the first transcontinental message was sent from California to New York and on to Washington DC: "May [this line] be a bond of perpetuity between the states of the Atlantic and those of the Pacific." The conciliatory message of 1861 seemed like a hopeful plea compared to Morse's 1844 proclamation of what God hath wrought. With the Civil War raging, Oliver Wendell Holmes wrote that the "first and obvious difference" between this war and all previous ones was that "the whole nation is penetrated by the ramifications of a network of iron nerves which flash sensation and volition backward and forward to and from towns and provinces."[38] Historian Edward Ayers confirms Holmes's view: "A long brewing sectional animosity boiled over when railroads, telegraphs, and newspapers proliferated in the 1840s and 1850s."[39] In addition to stoking enmity between regions and states, the telegraph influenced the results. A "wired" war allowed newspapers to send updates from the battlefield that could then be passed to the waiting public. Commanders telegraphed reports to headquarters and to facilitate the movement of troops. Almost fifteen thousand miles of telegraph lines were built to help fight the Civil War, and most of them were abandoned soon after, adding another layer to the devastation that the war inflicted on the American landscape.

The development of the telegraph industry coincided with a period of massive expansion in terms of the nation's size and a consolidation of wealth and management. When Morse inaugurated the telegraph, approximately forty thousand dollars had been invested in a forty-mile line. Missouri was the most western of the twenty-six states. In 1866 Congress passed the National Telegraph Act in an attempt to regulate the siting of lines, to promote rivalry between telegraph companies, and to limit the power of Western Union's burgeoning monopoly.[40] By 1867 the capitalization of the American telegraph industry had grown to forty million dollars, and eleven more states had officially entered the republic.[41] Many of these new states—Texas, California, Nevada, Kansas, Iowa, Wisconsin, Nebraska— were larger in terms of acreage than any of the original twenty-six. Telegraph lines were required to maintain bureaucratic control over these disparate regions. The visions of a "vast framework" that might be charted on maps corresponded with visible changes in the countryside.

Again, the railroad looms largest in the American narratives of nineteenth-century development and progress. The steam engine, the steel rails, the train bridge, the water tank, and the station platform, remain crucial industrial icons. In the 1850s and 1860s, however, the telegraph lines and railroad became inseparable, from visual and system standpoints. The telegraph provided the "most important technological addition to the railways," according to historian Wolfgang Schivelbusch.[42] Train dispatchers relied on the telegraph lines that ran parallel to the tracks to signal ahead to avoid dangerous head-on collisions and to keep a consistent schedule based on nationally regulated railroad time.

The telegraph lines also framed the train travel experience. In 1851 one passenger described hearing "the pulse of the engine throbbing quicker and quicker, and the telegraph posts seem to have started off into a frantic gallopade along the line."[43] With the addition of telegraph poles alongside the rails, Schivelbusch explains, "no longer did [the traveler] see only the landscape through which he journeyed, but also, continuously, the poles and wires that belonged to the railroad as intimately as the rails themselves do. The landscape appeared *behind* the telegraph poles and wires; it was seen *through* them."[44] Rail travel offered Americans the first glimpses of what millions of Americans now ride or drive by everyday—the irregular, repeated rectangle made by parallel posts (the rectangle's sides), the concatenated line or lines on top, and the horizon on the bottom. As visual artifact, the telegraph was the railroad's sinewy sidekick; as a practice, however, according to Daniel Czitrom, telegraphy "split communication (of information, thought) from transportation (of people, materials)."[45]

This split distinguishes telegraph from other revolutionary communication technologies. The spread of ancient orthographic systems and the invention of the movable type printing press in the fifteenth century also fundamentally altered the creation, distribution, and preservation of words and ideas. The telegraph, like previous innovations, circulated messages farther and faster than ever before. In the late 1850s the Pony Expresses that carried mail and packages across the Great Plains shaved ten days off the time it took messages to reach the Pacific coast. Within days of the inauguration of the transcontinental line in 1861, the Pony Express was rendered obsolete and ceased operations. With the correct connections, telegraphed messages annihilated space and time compared to messengers or shipments powered by foot, by horse, by boat, or by locomotive machine.

The speed and scope of telegraph networks increased the speed of written correspondence and capacity of the printing press. The telegraph could send information to many points and collect replies in a single hub. By the 1860s single wires carried multiples messages in different directions. The telegraph, printing press, and railroad worked together. Breaking news in any well-connected city could sweep across other urban areas in a matter of minutes. The advent of the steam press further shrunk the time required to send, receive, and print a story for mass distribution. With the development of railroad systems, printed reactions to the latest news could also be distributed more widely.

In each of these scenarios the telegraph initiated faster and broader circulation of information; however, telegraphy did not require anything to be "written." The invisibility of telegraph messages and other coded electric impulses makes it seem, as Czitrom suggests, that the information or thoughts contained in those messages had been divorced from material components (such as paper). Yet telegraphy—at least until the advent of wireless—required wires. The view of telegraphy as a dematerialized or disembodied practice stems from the physical qualities of electricity as well as the fact that only a small sliver of the population understood and touched its infrastructure. Line workers maintained the wires, poles, insulators, and relays. Another group of professional operators sent and received the messages. Considering the tariffs imposed on even short telegrams, it stands to reason that only wealthier individuals and representatives of larger business operations such as transportation, finance, and journalism regularly sent and received direct messages.

Telegraph *lines* seemed to resist the split between thought and material. Relatively few Americans directly engaged the telegraph as they might have engaged with other machines such as the rifle or the railroad. Instead, as David Hochfelder explains, most Americans "encountered the telegraph as a source of information in their daily newspapers."[46] The lines in one's environment provided another important means for encountering the telegraph. Indeed, without sending or receiving a message or reading a telegraph-sourced story in the newspaper, many Americans could see the telegraph as a visual artifact upon the landscape.

The Line as Industrial Icon

The telegraph line's appearance in mid-nineteenth-century visual culture, especially popular prints, sketches, and landscape paintings, reinforced its position as an icon of industrialization and national cohesion. Two famous images utilize telegraph lines to send a similar, albeit ambiguous message: "progress."

In Asher Durand's *Progress: The Advancement of Civilization* (1853) the telegraph line controls the movement of the eye across the canvas (fig. 5). The telegraph starts in the middle foreground, climbs up the right side of the frame, dips toward the midway point, and then disappears. As the line descends the hill and winds first to the right and then back to the left, the poles seem increasingly smaller, and by the time they reach the valley, it is difficult to distinguish the poles from the fence posts that surround the

5. The icons of technological advancement along the right half of the canvas—steamboat, railroad, log cabin, and wagon trains—conclude with telegraph poles that may have been harvested from where Native Americans stand observing from a ridge surrounded by recently felled trees. Asher B. Durand, *Progress, or The Advance of Civilization*, 1853. Oil on canvas, 48 x 72 in. Privately owned, Metropolitan Museum of Art.

cow pasture. The train tracks also move from the bottom right to left and meet the locomotive and its billowing smoke in right center. The tracks, in tandem with the telegraph line, draw attention to the lake, where steamboats are docked next to factories and mills. The industrial area emits its own puffy clouds and serves as the landscape's vanishing point.

The frontier icons—the wagon, log cabin, train, bridge, railroad, and steamboat—support the dominant narrative about settlement and industrialization. When Durand's painting was first exhibited, a review in the *Knickerbocker* magazine announced: "[Durand's *Progress*] is purely AMERICAN. It tells an American story out of American facts, portrayed with true American feeling, by a devoted and earnest student of Nature."[47] If the image is first and foremost "American," then the depicted technologies and their movement across the landscape—including the displacement of those already living there—seem like preordained facts.

Durand was a devotee of nature, but he also lamented the destruction wrought by American industrialization. In "Letters on Landscape Painting" he advises the young American artist to seek the "forms of Nature yet spared from the pollution of civilization." The placement of Native Americans in *Progress* seems to hint at the ramifications of civilization. This group, apparently vanquished by the advancing settlers, stand amid trees that have been cut to the stump. The forest has been recently cleared, and some of the timber may have been used for the telegraph poles depicted to the right. In that corner the dirt road, log cabin, horse-drawn wagon, and telegraph line represent the present, action, and the first push of so-called progress. The present moves toward the future, following the line of industrial machines until it reaches the mill and the image's apex—a brilliant sunset. The three sections form a visual and conceptual triangle of native past, frontier present, and industrial future. The viewer, like the group of Native Americans, gazes from the platform of an untamed past toward a murky present and into a pastoral future. Similarly, the telegraph line, with its scraggly, leaning poles and bending route, provide a subtle link between past, present, and future.

The telegraph line in John Gast's *American Progress* (1872) sends a more heavy-handed message (fig. 6). Here the telegraph line is dropped like spider silk from the hand of an angelic, scantily clad messenger. Gast's painting reinforces manifest destiny ideologies. The explorers and fur traders chase away the Native Americans and the buffalo herd. Behind them are the farmers, wagon trains, and railroads. A diaphanous figure floats above this chain of allegorical symbols and events. She carries a book of education in one hand and a spool of telegraph wire in the other. The telegraph is laced over the empty landscape in a visual pattern that attempted to reflect how peoples and technologies moved westward. It is worth noting that, in the allegory, the telegraph precedes the railroad. In many areas of the Great Plains and the western United States, the first infrastructure to appear on the landscape was a telegraph line.

A figure such as Gast's angelic messenger also appears in Alexander Jones's *Historical Sketch of the Electric Telegraph* (1852), a book that, in less

6. Tour guide and publisher George Crofutt commissioned Brooklyn-based John Gast to illustrate the "grand drama of Progress." In *Crofutt's New Overland Tourist and Pacific Coast Guide* (1878–79) he explains that the angelic figure bearing the "Star of Empire" on her forehead "carries a book—common school—the emblem of education and the testimonial of our national enlightenment, while with the left she unfolds and stretches the slender wires of the telegraph, that are to flash intelligence throughout the land." John Gast, *American Progress*, 1872. Oil on canvas, 12¾" x 16¾," Autry National Center, Los Angeles.

than two hundred pages, summarizes the progress in electrical science that led to Morse's device, reviews the latest technical improvements, and lists current stations and tariffs. (It notes, for instance, that a ten-word dispatch from Winnebago, Wisconsin, cost two dollars.) For much of the mid-nineteenth century Jones's book was "cited as authority on matters relative to the telegraph and telegraphing."[48] In addition to providing technical descriptions, Jones poeticizes the telegraph line. The epigraph quotes act 2, scene 1, of Shakespeare's *Tempest*, when shrewd sprite Robin (also called Puck) boasts, "I'll put a girdle round the earth in forty minutes." In the context of the play Puck has agreed fly around the earth to retrieve a magical purple flower, "love-in-idleness." Unlike Puck, the telegraph could not, in 1850, "girdle" the earth, but the pace of development had already instilled the idea that information would soon travel across the entire globe with supernatural speed (fig. 7).

In Jones's introduction the allegorical figure seems to sweep across the continent: "In one moment we find it [the telegraph] conveying messages of intelligence in advance of time over a continent, measuring the degrees of longitude, and dropping copies of its news at each hamlet, village and town in its flights over mountain peaks 'Where Alpine solitudes extend' across valleys wide and rivers deep and strong; and as quickly at its post again."[49]

The telegraph helped surveyors and cartographers make more exact measurements of longitude. The use of open circuits also made it possible for the news sent between major hubs like New York, Washington DC, and New Orleans to be "dropped" at all of the stations in between. In Jones's passage telegraphic communication is so fast that it troubles temporal structures: messages fly across the continent both "in one moment" and "in advance of time." A similar notion that telegraph wires carry messages faster than thought appears in Benjamin French's 1845 poem "The Changes of the World": "Swifter than thought th' intense and subtle fire / to do man's bidding flies along the wire."[50] In the final phrase Jones (perhaps deliberately) misquotes Oliver Goldsmith's poem "The Traveler" (1764), replacing "where Alpine solitudes *ascend*" with "where Alpine solitudes extend." The change implies the linear ex-*tension* of wires over mountain ranges and across territories such as Nebraska and the Dakotas. The telegraph line delivers thoughts ahead of time and pulls the seemingly boundless frontier within reach of eastern settlements.

In contrast to this poetic rendering, Laurence Turnbull's *Electro Magnetic Telegraph: With an Historical Account of Its Rise, Progress, and Present Condition* (1853) presents the telegraph lines as objects that spark scientific inquiry. "No one," he explains, "can view the extensive lines, and hear of and see its wonderful, nay, magical effects, without a strong desire to become better informed of its history and mode of operation." The line had become a familiar feature of the landscape—"its lines of iron wire pass

HISTORICAL SKETCH

OF THE

ELECTRIC TELEGRAPH:

INCLUDING ITS

RISE AND PROGRESS IN THE UNITED STATES.

BY

ALEXANDER JONES.

"I'll put a girdle round about the earth in forty minutes."– SHAKSPEARE.

NEW-YORK:
GEORGE P. PUTNAM, 10 PARK PLACE.
M.DCCC.LII.

7. "I'll put a girdle round about the earth in forty minutes" is spoken by the mischievous fairy Puck in Shakespeare's *Midsummer Night's Dream*. Shakespeare uses *girdle* in the metaphorical sense, yet in the 1850s it seemed possible that telegraph wires would girdle the earth and allow for fast and efficient communication. Frontispiece to *Historical Sketch of the Electric Telegraph: Including Its Rise and Progress in the United States* (New York: George P. Putnam, 1852).

before their very doors and extend even into the most distant wilds of the country."[51] While many viewers recognized the lines by their rows of poles and signature shape, their function remained a mystery.

Turnbull's history of electrically transmitted messages begins with electrical experiments conducted in Germany, France, and Spain in the late eighteenth century. In 1798 a twenty-six-mile telegraph wire connected the capital of Madrid to the Royal Palace at Aranjuez. The line provided Infante Don Antonia daily updates.[52] Turnbull also describes and praises advancements such as Royal Earl House's printing telegraph. House's device had twenty-eight keys set up like a piano and allowed senders to type the letters of the Roman alphabet. The typed letters would be printed at the receiving end. In the 1850s House lines were built from New York to Philadelphia to Washington DC, but Morse's system was cheaper and more successful.

For a history book written in the first years of the telegraph's development, Turnbull's work is impressive. One reviewer praised the "plain, perspicuous style" with which Turnbull provided "the most complete and satisfactory work on Electric Telegraphs published in this country."[53] Although Turnbull and other scientists and scholars helped to shed light on some of telegraphy's "magical effects," many Americans still felt leery in their presence.

Web Lines

The telegraph line's role in mid-nineteenth-century spiritualism, poetry, and fiction suggests a different inflection than the visual iconography of landscape paintings or hagiographic tone of history books. In these more imaginative renderings, the line is like a spider web or a living nerve. Those within reach of the lines are connected, for better or worse, to all other minds, bodies, peoples, and nations in the network.

Electricity's importance to some of these authors, such as Hawthorne, Melville, and Thoreau, has been established. The relationships between the authors' depictions of telegraph lines has not yet been examined. In addition to the mysterious qualities of electricity, the line passing through the forest or strung alongside the train tracks symbolized a single thread in a broader meshwork buzzing with affect and intelligence. The telegraph line inspired and shaped visual, fictional, and theoretic engagements with landscape.

The climactic ending of Nathaniel Hawthorne's *House of Seven Gables* (1851) most clearly displays the telegraph line's central role in the industrialized landscape of the nineteenth century. The narrative action rises when Clifford and Hepzibah Pyncheon find the corpse of their cousin, Judge Jaffrey Pyncheon. Due to entrenched family feuds, the brother and sister fear they will be held responsible for Jaffrey's mysterious death and decide to flee the scene. For decades prior to their flight, Clifford and Hepzibah had kept themselves insulated within the House of Seven Gables, and thus, when they emerge into the street of this small New England town and decide to board a train, they feel suddenly "drawn into the great current of human life" and "swept away with it."[54]

Hepzibah's earlier manic spell reveals "all strong feeling is electric," and now, as the train begins to move, the agitated Clifford strikes up a conversation with a fellow passenger about the wonders of electricity: "the demon, the angel, the mighty physical power, the all-pervading intelligence!" Clifford proclaims: "Is it a fact—or have I dreamt it—that, by means of electricity, the world of matter has become a great nerve, vibrating thousands of miles in a breathless point of time? Rather, the round globe is a vast head, a brain, instinct with intelligence! Or, shall we say, it is itself a thought, nothing but a thought, and no longer the substance which we deemed it!"[55] Clifford's speech hints at the "spiritual telegraph," a common analogy of the age. Some Americans believed the telegraph mirrored another ephemeral network that circulated divine messages. The spiritual telegraph spanned space like a vast organism. When blended with the idea of the body politic, the organism connects Americans across the continent. As one newspaper in 1846 explained, the telegraph "makes the pulse at the extremity beat— throb for throb and in the instant—with that at its heart . . . In short, it will make the whole land one being—a touch upon any part will—like the wires—vibrate over all."[56] Keys or other machines that sent and received signals acted like sensory organs; wires provided the nervous system. The line's ability to serve as a "brain" excites Clifford.

The passenger assumes that Clifford has referred to the literal telegraph, and "glancing his eye toward its wire, alongside the rail track," the man replies: "It is an excellent thing; that is of course, if the speculators in cotton and politics don't get possession of it. A great thing indeed, Sir; particularly as regards the detection of bank-robbers and murders!"[57] The passenger has an equivocal reaction to the presence of the telegraph line. Brokers will likely use the lines to manipulate the cotton markets, yet in the hands of law enforcement, the line can help to maintain social order.

The possibility that he might be tracked by law enforcement unsettles Clifford. "An almost spiritual medium" like the telegraph, Clifford laments, should be "consecrated to high, deep, joyful, and holy missions. Lovers, day by day—hour by hour, if so often moved to do it—might send their heart throbs from Maine to Florida."[58] Rather than hunt criminals, the lines should throb with affect.

Hawthorne's scene embodies the conflicting readings of the telegraph line as an emotional conduit and an industrial tool. Emerson said that machines in the landscape can be divine, but they "are not yet consecrated in their reading."[59] Clifford echoes this call when he says that the "great nerve" should be "consecrated to high, deep, joyful, and holy missions."[60] Clifford believes that electricity will change the material world until it is "no longer the substance which we deemed it."[61] This is a kind of naive social determinism: If the telegraph is only used for "holy missions," then the telegraph line can be considered as divine as the electricity it carries. The fellow passenger, and Hawthorne, knew that the telegraph was used for practical, and sometimes devious, purposes. Spiritual feelings may be like electricity, but "consecrating" the telegraph as a spiritual medium ignores the crony capitalism and rampant industrialization that facilitated the development of wire networks.

Published the same year, Herman Melville's *Moby-Dick* (1851) also employs a metaphorical telegraph wire to represent the uncertain trans-mission of emotions as well as the occupation (and exploitation) of the landscape. First, Ishmael references a telegraph wire in his distinction between "Fast-Fish" and "Loose-Fish": "Alive or dead a fish is technically fast, when it is connected with an occupied ship or boat, by any medium at all controllable by the occupant or occupants,—a mast, an oar, a nine inch cable, a telegraph wire, or a strand of cobweb, it is all the same." Any object that extends from an occupied ship to a fish can suffice to make that fish "fast." Using a cobweb or a telegraph wire to hunt a whale is unrealistic, but the passage offers another example of Melville's engagement with contemporary symbols and events. The telegraph wire here symbolizes connection across nations (or ships), control of prey, and mark of ownership. This line-as-stake relates to the overall didactic function of the chapter, which concludes with Ishmael asking: "What was America in 1492 but a Loose-Fish, in which Columbus struck the Spanish standard by way of waifing it for his royal master and mistress? . . . What at last will Mexico be to the United States?"[62]

Ishmael views the Fast-Loose dialectic governing language, politics, religion, and even "the great globe itself." Nations attempt to occupy and control other countries, writers attempt to fasten thoughts to words and sentences, politicians attempt to lasso support for their agendas, and the titans of industry attempt to harness the "loose" landscape. But these things (nations, voters, readers, landscapes) are Fast-Fish *and* Loose-Fish, too, because even if they are staked by certain attitudes, ideologies, or "lines," the hunters of the world (be they armies, ideas, or institutions) will continue their assault. Anything might be fastened, or refastened. Melville's inclusion of a telegraph line in this extended metaphor suggests that in the first five years of its development the line already evoked tensions between colonialism, industrialization, and any loose parts of the planet.

The second telegraph wire in *Moby-Dick* evokes the lines' power to metaphorically transmit affect. When Ishmael's crew harpoons a whale, the animal dives deep below the surface. The men wait in the boat until Starbuck cries, "Stand by, men; he stirs!" They sit bobbing in the boat, waiting until "the three lines suddenly vibrated in the water, distinctly conducting upwards to them, as by magnetic wires, the life and death throbs of the whale, so that every oarsman felt them in his seat."[63] The slight jerks and vibrations conduct intelligible messages, or "throbs," across the wire between animal and man.

With telegraph networks still in their infancy, Hawthorne and Melville were able to read between the lines, so to speak. They each critiqued the dominant views of this new advancement in communication. They accepted that telegraph lines may transmit throbs between bodies and across the globe, but the lines also wrapped anyone seeking independence or autonomy into an expansive and uncertain web.

Line as Harp

In the summer of 1845, as the first slivers of telegraph wire began to reach up and down the Eastern Seaboard from Washington DC, the twenty-seven-year-old Henry David Thoreau built himself a one-room cabin next to Walden Pond with the intention "to live deliberately, to front only the essential facts of life." The experiment concluded almost two years later, in 1847, and serious revisions of the journal he kept while living at the cabin began in 1851, the same year that telegraph lines reached Concord. In Thoreau's famous rebuke of the telegraph, he declared, "We are in great haste to construct a magnetic telegraph from Maine to Texas; but Maine and Texas, it may be, have nothing important to communicate."[64] Between 1846, when the United States went to war with Mexico, and 1854, when the Nebraska-Kansas Act declared that the new territories could decide whether or not to allow slavery, the farthest parts of the nation did, in fact, have important things to communicate. Most Americans agreed that telegraph lines from north to south and east to west kept open useful, even necessary, lines of commerce and communication. Thoreau, meanwhile, worried that "construct[ing] a magnetic telegraph [line]" and using it to "talk fast, and not to talk sensibly," would clutter minds with useless news and eventually reap harsh consequences.

Thoreau's negative views of telegraphy match the sentiments of Hawthorne and Melville. Thoreau's critiques support the popular construct of him as a romantic, hermetic Luddite who prophesied modern laments about social media—emoticons, twitter rants, and sensationalized Facebook posts seem like means of talking fast and not sensibly. However, a comparison of Thoreau's journal entries and the final publication of *Walden* in 1854 prove that while Thoreau criticized the messages commonly sent over the telegraph wires, he embraced "telegraphing" the fleeting experiences of nature.

In *Walden* Thoreau recounts: "So many autumn, ay, and wintery days, spent outside the town, trying to hear what was in the wind, to hear and carry it express! . . . At other times watching from the observatory of some cliff or tree, *to telegraph* any new arrival."[65] Thoreau's use of the verb *to telegraph* likely refers to an older form of communication—the semaphore, or "optical telegraph," an example of which Thoreau saw in 1851 on the coastline outside Boston and wrote about in his journal.[66] Even after the introduction of the electromagnetic telegraph, American ports like the ones near Boston still used semaphore telegraphs to signal to ships at sea. The semaphore also seems to inform Thoreau's advice to "be your own telegraph, unweariedly sweeping the horizon, speaking all passing vessels bound coastwise."[67] The telegraph sweeps outward like a lighthouse beacon; its messages are not directed *to* anyone (or anything) in particular. For Thoreau to "telegraph" is to send a message into the surrounding environment like a lightning bolt through an electrostatic field. This self-owned and self-operated telegraph is representative of his overall philosophy: true freedom requires control of oneself, one's labor, and one's modes and purposes of communication.

References to telegraphy and telegraphing in *Walden* pale in comparison with the repeated, unbound amazement Thoreau expressed for the mystical "telegraph harp." His first reference to the telegraph harp appears in *A Week on the Concord and Merrimack Rivers* (1848), which was drafted in earnest during his stay at Walden Pond and which recounts an 1839 canoe trip with his brother, John. In a moment of apparent anachronism, Thoreau says as he and John walked near the railroad in Plaistow, New Hampshire, they heard "a faint music in the air like an Aeolian harp, which I immediately expected to proceed from the chord of the telegraph vibrating in the just-awakening morning wind." In what would become a regular habit, Thoreau says he pressed his ear to the wooden post and heard a celestial message, "sent not by men but by gods." Thoreau related the humming sound to "the first seashell heard on the seashore," and he found meaning in this archetypal white noise: "It told of things worthy to hear, and worthy of the electric fluid to carry the news of, not of the price of cotton or flour, but it hinted at the price of the world itself and of things which are priceless, of absolute truth and beauty."[68] Of course, Thoreau did not likely hear the electromagnetic telegraph during his trip with his brother in 1839, but later journal entries prove he repeatedly stood next to telegraph poles and waited for the transcendent tunes.

Between 1851 and 1861 Thoreau made over thirty references to the telegraph harp in his journal. His first report questions the new technology's effectiveness: "In a day or two the first message will be conveyed or transmitted over the magnetic telegraph through this town, as a thought traverses space, and no citizen of the town shall be aware of it. The atmosphere is full of telegraphs equally unobserved. We are not confined to Morse's or House's or [Alexander] Bain's line."[69] Like the other technologies he criticizes in *Walden*, telegraphy seems to limit the individual's search for higher, nobler messages. The next day Thoreau went to inspect the line himself. The line was strung on stripped wooden poles next to the train tracks near a ravine named the Deep Cut. He describes the first surprising encounter in his journal: "As I went under the new telegraph wire, I heard it vibrating like a harp high over head. It was as the sound of a far-off glorious life, a supernal life, which came down to us, and vibrated the lattice-work of this life of ours."[70]

Other Americans also imagined telegraph lines as the strings of a mystical instrument. In 1846 S. M. Partridge described telegraph wires as "strings on the fiddle of Animal Magnetism; and not till they are stretched from pole to pole can a perfect tune be expected; then will it raise its immense fiddle-bow wide as the heavens, and strike a harmonious movement that shall rock all the earth into a slumber."[71] Similarly, in the short story "An Evening with the Telegraph Wires," the protagonist climbs a tree, touches a telegraph wire, and discovers that he can listen to messages being sent through the lines and feel the sender's emotions. The discovery sends the narrator into a series of ecstatic reveries, including the "gigantic fancy" that "this State of New York is a great guitar; yonder, at Albany, are the legislative pegs and screws; down there in Manhattan Island is the great sounding-board; these iron wires are the strings!" Contact with this guitar reveals the "invisible wires that connect one heart with another."[72]

The metaphorical fiddle or guitar may have played sweet chords, but the real telegraph harp often made experimental noise. William Bender Wilson recalled that the telegraph wires in rural Pennsylvania made "musical, weird, fantastic sounds" that startled locals: "The public mind having something of a superstitious bend . . . would walk a very considerable distance out of their way . . . to avoid passing under or near it."[73] An 1847 report from Newcastle, England, said the wind over wires produced a "beautiful tone . . . resembling that of a Aeolian harp,"[74] but more often the wires emitted a high-pitched, eerie whistle. Thoreau found that sound intoxicating. He described it as "a beautiful paucity of communication" and "the finest strain that a human ear can hear." The whispering signals were a "triumphant though transient exhibition of truth."[75]

Thoreau took detailed notes about "this magic medium of communication for mankind!" as if he were preparing a field study. He noticed that the harp responded differently to heat and cold. He placed his ear near the wooden posts, "where the vibration is apparently more rapid." He also observed that the telegraph harp "does not require a strong wind to wake its strings; it depends more on its direction and the tension of the wire."[76] Thoreau spent the fall of 1851 making his inspection and trying to attune his mind and body to the music. The next spring the harp helped to inspire one of the most climactic passages in American literature.

On March 9, 1852, Thoreau writes in his journal about his daily walk through the Deep Cut, a land owned by the Fitchburg Railroad. The Deep Cut appears often in Thoreau's journals and in *Walden*, serving as a kind of dividing line between the wilderness of Walden Pond and the civilization of Concord. On March 9 Thoreau begins his journal entry by describing the thawing soil on the banks of the railroad, which is "perhaps our pleasantest and wildest road." From the rails below and banks on either side, Thoreau's scope rises to include the telegraph harp above: "When I hear the telegraph harp, I think I must read the Greek poets. This sound is like a brighter color, red, or blue, or green, where all was dull white or black. It prophesies finer senses, a finer life, a golden age. It is the poetry of the railroad, the heroic and poetic thoughts which the Irish laborers had at their toil now got expression,—that which has made the world mad so long. Or is it the gods expressing their delight at this invention?" The telegraph harp sings "the poetry of the railroad" and thus might be the aspect of industrialization viewed most favorably by the gods. The sounds made by the telegraph harp eventually turn Thoreau's attention back to the eroding soil: "The flowing sand bursts out through the snow and overflows it where no sand was to be seen. I see whether the banks have deposited great heaps, many cartloads, of clayey sand, as if they had relieved themselves of their winter's indigestions."[77] The description of the harps' song is enmeshed with the description of thawing banks. This journal entry informed a crucial moment in *Walden*.

In Thoreau's journal the sounds of the telegraph harp speak of "a finer life, a golden age." In the climatic chapter "Spring," Thoreau observes pulpy streams of sand and mud oozing from the Deep Cut. The image, for Thoreau, is "the realization of a Golden Age." The journal's lines about the eroding bank match those of the *Walden* passage, but Thoreau's book embellishes with a series of archetypal associations. The banks are lava-like and ooze "pulpy sprays" and the "laciniated lobed and imbricated thalluses" of lichens. Thoreau sees "coral, or leopards' paws or birds' feet, of brains or lungs or bowels, and excrements of all kinds." The sandy rivulets that pour through the banks overlap and interlace, and Thoreau's description seems to verge on hallucination when he notices that "this sandy overflow is something such a foliaceous mass as the vitals of the animal body."[78] *Walden*'s description of the thawing banks is crucial to the philosophical transformation that is "Spring"; however, the sounds of the telegraph harp accompanied the original experience upon which this pivotal passage was derived.

We can imagine Thoreau standing beneath the wire, turning his head toward the sky, closing his eyes, and dissolving like Emerson's eyeball. When the sound faded away, he might have opened his eyes and felt as if the powers around him were in balance. Thoreau, famous for his diatribes against technology and government, stood aside, awed and humbled by this strange and unintended effect of the nation's newest infrastructure. While *Walden* should still be read and taught as a sobering reminder of our perpetual attraction to the latest technological wonder, it is telling that Thoreau hoped we might attune ourselves to one particular part of our new technological landscape, the telegraph line, so that we might hear its organic, electric, and uplifting messages across time and space.

Wired Frontiers

During the Civil War and subsequent era of Reconstruction, as the first lines were erected in California and then the transcontinental telegraph connected the East (via Omaha) to the West (via Carson City, Nevada, and Sacramento, California), widespread telegraphic infrastructure developed into a critical, necessary network. With this increased dependency came more widespread skepticism and anxieties, especially in the West, where lines could be wielded as an instrument of colonization and, conversely, sabotaged or severed.

Some midcentury onlookers believed North America, a frontier of "wild beasts, and still wilder Indians," had suddenly been transformed and seemingly replaced by "frail wires" and other machines; however, the peoples already living in the Western Territories that eventually became Nebraska, Colorado, Wyoming, Utah, Nevada, and California initially engaged with the telegraph in ways similar to their Anglo counterparts. In other words, Native Americans reacted to wires with initial disbelief and wonder, followed by a basic understanding of the technology's advantages and fear of its broader implications. In the East the lines aroused suspicions about deceitful messages and shady finance; in the West the presence (or absence) of the lines had sharper, sometimes violent repercussions.

The builders of the first lines to cross the western United States took various approaches when introducing this new technology. James Gamble used relatively progressive methods. In 1852 Gamble was superintendent for the first major line through California, which was strung between San Francisco and Marysville via San Jose, Stockton, and Sacramento. According to Gamble, as the teams erected poles and nailed the crossarms, a "certain devout Mexican woman" believed they were crosses used to ward off evil. She exclaimed, "I believe those Americans are becoming good Catholics!" When the line was complete, Gamble gathered the locals into his telegraph office. He showed them a letter he had written and then made it "disappear" into the wires. His display was meant to suggest the basic premise of telegraphy, but it also led the people to believe the letter had traveled along, or through, the wire. Overall Gamble's practical joke seems to have been relatively harmless, and he nurtured amicable relationships with locals across the West.[79]

In another instance, in 1860, Gamble's partner, James Street, met the chief of the Nevada Shoshones, Sho-kup, and explained the project they were undertaking. Sho-kup called the telegraph, "We-ente-mo-ke-te-bope," or "wire rope express," and Street offered to let Sho-kup use it to send messages. Sho-kup declined the offer, but the respectful exchange provided a foundation for the friendly relations between telegraph workers and Shoshones during the construction of the transcontinental line. In 1861, as the line was nearing completion, Gamble hired a number of Native Americans both to care for the crew's livestock and to gather poles, a scarce commodity in the desert regions of Nevada and Utah. Gamble said the men were efficient workers, and he hoped "that they might report to the different tribes how well they were treated, and in this way favorably influence the Indians toward the members of the party and the telegraph line."[80] In Gamble's hands the telegraph line was mysterious but not malicious.

Conversely, Edward Creighton and his team, including Charles Brown, introduced the telegraph to the Sioux of Nebraska and Wyoming via shocks. As Brown recalls in his diary in August 1861, Creighton visited a group of Sioux near Fort Laramie and had an operator, W. B. Hibbard, bring along a battery and wire. Through a translator Creighton explained that it would be dangerous to interfere with the line's construction. To demonstrate, he had some of those present join hands and then had Hibbard release a shock. According to Brown, they looked upon Creighton as a "big medicine man."[81] In a separate, unverified incident, Creighton asked one chief to visit a station in Scottsbluff and another, possibly an Arapaho chief, to go to a station in Chimney Rock. Through interpreters the chiefs were asked to send messages to one another and then to meet at a place halfway between to compare the veracity of what each had said over the wire.[82] Captain Eugene Ware interpreted the meeting differently, saying: "No effort was made to explain it to the Indians upon any scientific principle, but it was given the appearance of a black and diabolical art. The Indians were given some electric shock; and every conceivable plan, to make them afraid of the wire . . . to make the Indian dread the wire."[83]

Eventually, such tactics backfired. The lines that passed through Nebraska, Colorado, and Wyoming would be repeatedly attacked and broken during various battles and raids in the 1860s. In 1865 one newspaper complained of the attacks by "large bands of marauding savages, who now know that they are inflicting injury by destroying the telegraph line . . . fully aware what a tender point it is."[84] It is impossible to know for sure how the different methods used to acquaint different tribes with overhead infrastructure influenced their subsequent reactions. However, it is telling that through vandalism and guerrilla warfare, Native Americans provided the first widespread resistance to overhead wires.

2

New York's Frontier Lines and Telegraph Forests, 1882-1916

At the turn of the twentieth century the state of New York exemplified the gulf between two competing readings of overhead infrastructure. New overhead power lines appeared as wondrous threads; others, such as the telegraph networks in Manhattan, appeared as malevolent webs. These opposing reactions to overhead lines were framed in frontier rhetoric and a long-standing moral dialectic—in the United States heroic and righteous explorers (and inventors) overpowered uncertain, "untamed" landscapes (and technical constraints). Frederick Jackson's Turner's "frontier thesis" and Nikola Tesla's "On Electricity" offer context for the advent of "pioneer" power lines; they also suggest why the untamed telegraph forests blighting urban areas were viewed as "wire evil."

New York's first long-distance power lines began operating in 1895, when a sliver of the 3,160 tons of water per second dropping over Niagara Falls was converted into alternating currents and sent to Buffalo. To build and connect long brick tunnels, metallic chutes, steel turbines, and various switches and transformers in the nation's first massive power plant required state-of-the-art engineering expertise. When completed, the spectacle at Niagara Falls revealed aspects of the natural sublime, technological sublime, and electrical sublime. Visitors witnessed the sublimity of electricity, Nye argues, as the powerful floodlights focused on the cascade and "dissolved the distinction between natural and artificial sites."[1] The wires, cables and wooden poles surrounding this "electrical Mecca of the world" were not featured in the same way as the lights and the waterfall, but they also astonished and terrified.[2] Articles such as "20,000 Horse Power over One Wire" hinted at the incredible feat.[3] A New York City art critic who visited the falls and the plant viewed this "force of Nature tamed and held by a silken thread" as "the most wonderful thing in the world."[4] Mentions of the wondrous power lines appeared in speeches, trade magazines, and newspaper articles.

In contrast to the all-encompassing transmission of Niagara power, the metallic nets in Manhattan were eyesores. By the 1880s thousands of miles of telegraph wire blanketed the city and stretched through every borough (fig. 8). Construction crews and linemen tacked wires and glass insulators to rooftops and stamped wooden poles into sidewalks. Overburdened poles on major avenues carried dozens and sometimes hundreds of telegraph and telephone lines owned by different companies. Haphazard and makeshift networks covered entire neighborhoods such as the West Side, where one could see ninety-foot poles carrying thirty crossarms and up to three hundred wires each. When lines for lighting and power began to appear in the streets, dark chaotic bands already crossed Gotham's skyline.

The conflicting responses to electricity and wired environments were central to two pulp novels published at that time, *The Wire Tappers* (1906) and its sequel, *Phantom Wires* (1907), by Arthur Stringer. In both novels wire networks symbolize the enthusiasm, and anxieties, of the so-called electrical age. Jim Durkin, a former telegraph operator who falls into a life of crime, sees the relatively unguarded wires as opportunities. He speaks excitedly about the "little condensed Niagara of power" outside a house in the Diamond district. He plans to "capture and tame and control" the current in the wire and "make it my slave and carry it along with me almost in my pocket, on a mere thread of copper."[5] His partner in crime and romance, Frances Candler, sees the physical wires as representative of dark, mysterious forces. She laments the presence of "ghostly wires connecting us with one another . . . our past, and with our ancestors, with all our forsaken sins and misdoings."[6] Hawthorne's character Clifford voiced concerns about single, frail telegraph line chasing criminals. Approximately fifty years later, Stringer's heroine suggests the wires are omnipresent and, with the advent of wireless technology, impossible to escape.

8. Telegraph and telephone wires in New York, ca. 1887. This view of Jones and Broadway is a few blocks from where Edison opened his Pearl Street Station. Library of Congress Prints and Photographs, LC-USZ62-75483. ©1940, Creative Educational Society. Used by permission of The Creative Company.

The literal and fictional engagement with electric lines in New York represents an important change in the collective perception of electric infrastructure. Other states and municipalities faced similar conflicts as the new lines required for telephony and then electric power joined the existing telegraph apparatus. The three types of lines often occupied the same rights-of-way and sometimes the same poles. Photographs from Muncie, Indiana, taken in 1897 show "major streets cluttered by five-tiered electric poles, with a maze of telephone, telegraph, and electric wires overhead."[7] If wires cluttered the streets of Muncie, they deluged New York, Philadelphia, Boston, and Washington DC. The hordes of wires set citizens on edge, such that proposals for new electric trolley systems that utilized overhead lines met with fierce public resistance. Communities desired public transportation but only if such systems did not add another layer of overhead lines. The trolley debates included competing visions of municipal financing, public transportation, and urban expansion, but "aesthetic arguments provided the strongest objections . . . almost everyone agreed that the wires were unattractive."[8] Expectedly, promoters, politicians, and business owners seemed willing to dismiss the aesthetic inconvenience. After all, lines for electric communication networks and then trolleys, factories, and street lighting helped their cities grow broader, wealthier, and more spectacular. Thus, the increasingly complex and wondrous advantages of electricity were disassociated from the material infrastructure that electrification required.

While the juxtaposition of responses to power lines and telegraph lines in New York clarifies the conceptual disconnect between infrastructure and its aesthetic impacts, a moment in Chicago, 800 miles west of New York City and 550 miles from Niagara Falls, provides the pivot between the equivocal reactions to overhead wires. This was a moment when the material and rhetorical constructs of electric technologies and American landscape were amplified and exchanged.

Electric Frontiers

On July 12, 1893, Frederick Jackson Turner delivered a paper titled "The Significance of the Frontier in American History" to a meeting of the American Historical Society at the Chicago Columbian Exposition. Shortly thereafter, the published text of Turner's address became a hallmark of American studies. Some historians have viewed it as "the most famous address ever delivered by an American historian" and "the single most influential piece of writing in the history of American history."[9] Turner's thesis still looms large, but by the 1950s historians such as Henry Nash Smith and Earl Pomeroy had forcefully disputed Turner's Anglocentric, male-dominated, romanticized frontier.[10] Subsequent scholars such as Patricia Nelson Limerick and Peggy Pascoe also challenged Turner's thesis by revisiting the crucial contribution of women, non–English speakers, and persons of color to western history. In addition, William Cronon's history of Chicago, *Nature's Metropolis* (1992), and Carroll Pursell's *Machine in America* (1995) show how eastern technologies and infrastructures controlled western development. Boosters in Chicago viewed the advancing frontier line less as a site of renewal or escape than as a way to reconfirm a "symbolic relationship between cities and their surrounding country sides."[11] The frontier process may have rejuvenated a sense of rugged individualism for some Americans, but it also solidified urban centers such as New York and Chicago as continental lynchpins of finance, transportation, and culture.

As a discipline, American studies has made crucial correctives to Turner's thesis and exposed its flawed conclusions; however, the timing of Turner's performance of the text helps to explain why his narrative about American history was so widely accepted. Turner and his fellow historians gathered to consider the frontier past, while around them engineers, electricians, and inventors were displaying an electrical future.

In Chicago that day Turner asked his audience to imagine something like an eighteenth-century time-lapse recording of migratory animals, hunters, traders, and farmers following one another in single file from east to west past the colonial line of the frontier: "Stand at Cumberland Gap and watch the procession of civilization, marching single file—the buffalo following the trail to the salt springs, the Indian, the fur-trader and hunter, the cattle-raiser, the pioneer farmer—and the frontier has passed by. Stand at South Pass in the Rockies a century later and see the same procession with wider intervals between."[12] Over the course of a century—between standing at Cumberland Gap (a mountain pass near the borders of Virginia, Kentucky, and Tennessee) and South Pass (in Wyoming)—the frontier line had swept from the Eastern Seaboard into the Great Plains by migration, by exploration, by trade, by massacre, by homesteading, and by treaties and land deeds (spanning the full spectrum of legality). The Louisiana Purchase, the discovery of precious metals in Colorado and California, and a steady procession of homesteaders, immigrants, and emigrants streamed toward the frontier and facilitated the settling of the West. Throughout the late eighteenth and early nineteenth centuries canoes, steamboats, and wagon trains flowed past that invisible frontier string. Wars, federal territories, and statehood filled in cartographic blanks until, in the decades after the Civil War, the American settlers had pushed to the Pacific and no clear frontier line remained. After three centuries of predominantly Anglo-European exploration and occupation, the seemingly endless continent seemed compassed, the howling wilderness mapped, and the millions of paths first formed by animals and foot traffic were further secured by telegraph threads, steel rails, and barbed wire. Finally, the frontier line disappeared. Indeed, Turner's argument begins by stating what for him was an alarming fact: in 1890 the U.S. Census Bureau reported that a "continuous line of frontier" was, for the first time in the history of Anglo occupation of the continent, no longer discernible on a map. As the line disappeared, nostalgia began to cloud popular views of "frontier" icons and lifestyles.

Turner did not define the physical frontier or provide a formula for its advance. Instead, he argued that more than politics, race, language, or technology, a unique frontier mentality defined the American character. Even after the frontier closed, the visceral attraction to frontier-like spaces and lifestyles would continue to shape what he called "the new product that is American."[13] Buffalo Bill's Wild West show, which had its breakout run during the summer of Turner's address, was the clearest and most attractive new frontier product in the United States. Buffalo Bill's troupe of Rough Riders and their Native American counterparts reenacted mythical (and racially prejudiced) scenes in an open area just outside the official fairgrounds. The show's success helped to initiate a distinctly American genre, the western.[14] The western, as live rodeo show and later as feature film, provided an avenue for many Americans to feel culturally connected to a collective past. Through westerns the idea of the frontier, as Turner suggested, maintained its influence in the American mind.

Turner also boasted that Americans would seek new physical frontiers. The need for physical fluidity and expansion—key components to the doctrine of manifest destiny—would not wane. "The American energy," Turner claimed, "will continually demand a wider field for its existence."[15] Beyond the continent the "wider field" could be located in various foreign policies and sites of occupation. For instance, the Monroe Doctrine, which stemmed from a State of the Union speech made by James Monroe in 1823, warned that any European attempt to manipulate or occupy an independent nation in North or South America would be viewed as "manifestation of unfriendly disposition" toward the United States. As the continental frontier line disappeared, the United States reinforced its commitment to protect, and control, neighboring countries, which had once been frontiers of European occupation. In addition, the expanding influence of the United States created some new frontiers—Hawaii, Alaska, Cuba, the Philippines, Puerto Rico, and Guam.

Turner romanticized a frontier line and its attractive force; Buffalo Bill's Wild West show romanticized a frontier lifestyle of horses, wagons, rifles, and other tools, which, from that point forward, became rustic emblems of the past. Meanwhile, electric lines staked out a wider field for the American future. The American energy that had been a driving force of excursions westward was reassigned to the technological development within the urban core.

America's New Frontier (1929), a small promotional text published by Samuel Insull's Middle West Utilities, makes an explicit argument for this transfer. Like Turner's thesis, *America's New Frontier* takes a sweeping and optimistic view of the "equalitarian" process of electrification. The authors suggest that in 1889, a year earlier than Turner suggested, "the march to free lands came to an end." Shortly thereafter, "a new frontier revealed itself—an internal frontier, consisting of great areas of neglected country between the cities." Electric power utilities "began weaving a web of power lines over this new frontier" to link villages and towns to the city and facilitate, in their view, "a more general diffusion of wealth and good living over the land." Despite it kernels of truth, it is clear that such promotional materials slanted more toward propaganda than accuracy. The spoils gained by taming the West or electrifying small towns were neither distributed equally nor designed to first and foremost facilitate the common good, as evidenced by the spectacular demise of Middle West Utilities, the Enron of its day. Investments and individualism, risk and reward, pushed the frontier line and, later, power lines across the landscape. The attractive force of a spatial frontier that evoked feelings of discovery and adventure was, in parts of the popular imagination, shifted to a new, technological frontier of comfort and interconnection.

To better understand that cultural shift, one might imagine visiting Chicago on the afternoon of Turner's address. Turner said his talk received a weak response and that not a single one of the approximately two hundred historians in attendance asked a follow-up question. One recent historian says this was likely because "most of [Turner's] audience had spent the day touring the 'White City'—the exposition ground on Chicago's South side—and many had accepted Buffalo Bill Cody's invitation for a special performance of his Wild West show."[16] Imagine witnessing the romanticized frontier performance and then listening to Turner speak about the frontier line of western settlement. As the day's scheduled events conclude and your colleagues go their separate ways, you stand at the center of Chicago's fairgrounds, on the Midway Plaisance.

The evening air cools, and the excitement builds. Couples and groups gather to get their first glimpses of electric light dropping down from poles like bright boughs on a moonlit tree. One advertisement for the fair promises, "The very ground on which the exposition is built . . . vibrate[s] to the hum of electrical machinery, and at night the air will . . . blaze with bright light springing into being through forces set at liberty by the subtle fluid."[17] You watch as electric trains, moving walkways, and escalators transport groups around the fair. Instead of riding, you walk, your eyes set on a massive structure in the distance. This novelty, designed by George Washington Gale Ferris Jr. to rival the Eiffel Tower, is one of the largest electric machines ever built. The 264-foot-tall structure, later named the "Ferris wheel," dominates the skyline. You approach slowly, gaze upward, and see carriages like white wooden sailboats rocking high above. Each carriage can hold up to sixty people; the entire bicycle wheel structure can hold up to a thousand. You hear laughter and exclamations as the cars rise. The wheel, turning on the largest piece of steel ever forged, lurches, ablaze in its own magnificent lights. You pay for a ticket, climb into a car, and feel the machine jerk forward. As you rise, you look out past the light bulb–studded rim of the Ferris wheel toward Lake Michigan, pondering the hidden network of tunnels carrying 104 miles of wire through the "Great White Way." The wires "placed underground because of safety and aesthetic reasons" cross beneath bazaars, shops, parks and buildings.[18] The lines bring alternating current (AC) from Machinery Hall to the thousands of lights and machines. The success of this specific installation has, incidentally, effectively ended the "Battle of the Currents" between the Tesla-Westinghouse system for AC and Thomas Edison's system for direct current (DC).

As the car rises 50, 100, and 150 feet into the air, you see crowds swarming down the Midway and flowing in and out of the sparkling facades of the White City. Figures move through shadows, and some fresh-faced tourists pause, lean their heads back, and bathe, amazed, in artificial light. The site represents 10 percent of the entire nation's artificial lighting—a significant chunk of American power concentrated into two square miles. In the distance massive searchlights beam through the night. An electric fountain spurts liquid jets bathed in red, yellow, and green on the black canvas. Just beyond the fountain, a queue snakes around the Electricity Building. Here the future is on display: Bell's telephone, Edison's phonograph and kinetoscope, Tesla's neon signs, and hundreds of other new electrical devices.[19] The spectacular light shows and revolutionary machines you see in Chicago that evening have set the standard for increasingly spectacular exhibitions at national expositions and international fairs that will be held in Omaha in 1899, Buffalo in 1901, and San Francisco in 1915. On this brilliant evening, after contemplating a vanquished frontier, you see the new one. You bear witness—you see America go electric.

During the exposition 21.5 million paid visitors (the actual number of visitors may have been closer to 27 million) are exposed to electric rides, electric gadgets, and displays of the electrical sublime.[20] Chicago provides glimpses of the past frontier lifestyle and visions of a future driven by the generation, transmission, and consumption of electricity.[21] The theme park–like, idealized vision of mass electrification did not, of course, include the banal intrusions made by overhead lines. The lines bringing electricity to the buildings, streetlights, and electrical rides had been buried in special conduits. Similar to Morse attempting to bury his telegraph line, the overhead maze was largely absent from the World's Fair. Other communities either could not or would not spare this expense. For the rest of the nation electrification brought a new wave of visible, overhead wires.

In the years between 1893 and 1915 every major city in the United States electrified. In the mid-twentieth century, after decades of efforts by the Rural Electrification Administration, one might say that the "electric frontier" had been closed. However, during the electrical age some of the most powerful lines of the electrical frontier—the frontier lines of long-distance transmission—were strung hundreds of miles to the east, in Upstate New York.

"On Electricity"

On January 12, 1897, at the Elliot Club in Buffalo, New York, hundreds of dignitaries gathered to celebrate what one newspaper called the "only electrical banquet the world has ever seen."[22] The event marked the completion of the nation's first massive hydroelectric power plant at Niagara Falls and the corresponding power lines that carried electricity across twenty-six miles to power Buffalo's streetlights, factories, and trains. The importance of transmitting electric power across great distances—at the time the twenty-six miles separating the Niagara and Buffalo seemed great—was a critical piece of the banquet's keynote, titled "On Electricity" and delivered by the Serbian-born scientist and inventor Nikola Tesla. Tesla had immigrated to the United States in 1882, his alternating current motor was patented in 1888, and throughout the 1890s he lived in premiere hotels in New York City and enjoyed celebrity status as "one of the most important persons living today."[23] In 1895 the *New York Times* stated, "To Tesla belongs the undisputed honor of being the man whose work made this Niagara enterprise possible."[24] The bronze patent plaque attached to the Niagara Power House listed Tesla's name thirteen times. While the twentieth century would arguably forget about Nikola Tesla, that night his contribution was fully recognized. The introduction of "the greatest electrician on earth" caused an uproar: "The guests sprang to their feet and wildly waved napkins and cheered for the famous scientist. It was three or four minutes before quiet prevailed."[25]

Tesla, who had severe social phobias, seemed overwhelmed by the attention. He admitted in his Eastern European accent that he felt so "full of the subject" that as soon as he attempted to express his ideas on electricity, "the fugitive conceptions will vanish, and I shall experience certain well known sensations of abandonment, chill and silence. I can see already your disappointed countenances and can read in them the painful regret of the mistake in your choice [of a speaker]."[26] Tesla was humbled and possibly terrified, but I believe his opening lines of self-deprecation should be read as a rhetorical gesture. He maintained control, and his "fugitive conceptions" soon became luminously crystallized decrees.

Just as the internet seems to have become an exceptionally strong force in the modern mind, Tesla prophesied that all great human achievements from that moment forward would require electric power. He explained that new electrical instruments would refine the exactness of knowledge, electrical science would increase the understanding of invisible forces, and electrification would enhance the "influence of the artist," who could perceive a single truth and be "consumed by the sacred fire."[27]

As his speech gained momentum, Tesla spoke on electricity *and* with it. Electricity was not just a potent factor of intellectual discourse or a tool of industrial progress; it was *the* characterizing feature of the mind and *the* gauge for progress. In Tesla's vision this pervasive, abundant commodity would be transmitted from hydroelectric plants placed at waterfalls across North and South America both to decrease dependence on limited fossil fuels and to make everyday life more comfortable. Worldwide electrification would erase ideological boundaries and spark new lines of inquiry across trades and disciplines; it would cure health problems, eradicate poverty, and end military conflicts. Lawmakers, economists, and philanthropists could only provide temporary relief for struggling masses. Tesla grandly stated, "The greatest significance for the comfort and welfare, not to say for the existence, of mankind . . . is the electrical *transmission* of power."[28] The problem of power generation, Tesla believed, had already been solved. Now he set his sights on how to transmit and distribute electricity to the rest of humankind.

Bombastic claims about the sweeping, positive effects of national and even global electrification were not uncommon in the 1890s. Tesla himself made a number of outrageous predictions about providing free, universal access to electricity. But "On Electricity" is remarkable for at least two more reasons. First, Tesla uses frontier rhetoric to describe the process of electrification, saying that the Niagara Power project was made possible by "intrepid pioneers."[29] After the 1890s *pioneer* was often used to describe scientists, inventors, and entrepreneurs. Edward Dean Adams's memoir *Niagara Power* (1927), for instance, frequently references the "pioneer" aspects of the project, including the "pioneer lines" that carried power from the plant. Each of the reflections submitted to Adams about the project also include the word *pioneer*: George Forbes, "engineers . . . all over the world . . . copy your *pioneer*, and therefore experimental, scheme"; Clemens Herschel, "Niagara Falls . . . a notable *pioneer* and example;" Harold Winthrop Buck, "Another matter in which the Niagara enterprise was a *pioneer* was in the large scale distribution of power by underground cables;" Paul M. Lincoln, "Niagara was not only a *pioneer* in the adoption of alternating current polyphase system on a large scale, but it was also a *pioneer* in many of the specific methods, practices, and devices that were essential to the ultimate success of that system."[30] The pioneer label and frontier rhetoric was attached to specific infrastructure projects as well as the more general pursuit of scientific knowledge. For instance, the attractive field of electrical science and what we might now call physics, Tesla explained, was the "unexplored, almost virgin, region, where, like in a silent forest, a thousand voices respond to every call."[31] Relating the world's most powerful application of science and engineering to a virgin forest settled by pioneers situates the Niagara enterprise as a continuation, and possibly a culmination, of the broader American project to explore, celebrate, and then tame the American wilderness.

Tesla's speech is also important in that it signals his turn toward the *wireless* transmission of power. In 1895, before the Niagara Power plant was complete, Tesla promised that he could "place 100,000 horsepower on a wire and send it 450 miles in one direction to New York, the Metropolis of the East, and 500 miles in the other direction to Chicago, the Metropolis of the West."[32] In the coming decades other electrical engineers realized Tesla's vision of long-distance power transmission. However, just as Morse imagined his own telegraph infrastructure would be channeled underground and out of sight, Tesla implied that power lines, overhead and underground, would not be necessary. He told the audience gathered in Buffalo that he would soon transmit power without connecting wires. In 1900, as Tesla prepared to launch his Worldwide Wireless project at Wardenclyffe Laboratory in Long Island, he claimed that Niagara Falls would "some day supply New York City, giving it all the electric power it can consume without the use of wires."[33] It is simultaneously peculiar and fitting that Tesla chose an event filled with engineers and financiers who had taken significant risks to build the first famous long-distance transmission line to assert that his new venture would remove the need for such lines.

Although Tesla did not accurately predict the success of wireless transmission or a world without power lines, his AC system unquestionably paved the way for the Niagara system, which helped standardize the grid and unleash electrification's revolutionary impacts. Tesla, unwittingly, helped power lines range across the Niagara frontier. Like Morse's telegraph lines, these power lines became icons of the era. We may wonder what the world might look like if Tesla had realized his wireless visions, yet it almost seems fortuitous that the earth's surface was webbed with wires; without Tesla's AC system, it might have been covered by tubes.

The Niagara Frontier: From Pneumatic Tubes to Sublime Lines

Niagara Falls is one of the most long-standing and preeminent sites of American wilderness and the American sublime. In 1836 Thomas Cole boasted: "Niagara! That wonder of the world!—where the sublime and beautiful are bound together in an indissoluble chain."[34] Beginning in the eighteenth century, the thunderous torrent was extolled by painters, poets, novelists, nature lovers, and eventually tour guides. Niagara aroused feelings of patriotism, aesthetic bliss, and moral catharsis. Above all, Niagara shocked its visitors. Guidebooks and paintings could not prepare a person for the stunning moment when one peered over the rim and confronted the powerful deluge.

To help preserve and protect that natural force, in 1885 the New York State legislature approved a proposal to create the Niagara Falls Reservation (now the Niagara Falls State Park). The legal maneuver helped to clear away the chaotic collection of water-powered mills and gaudy tourist traps around the cataract. Frederick Law Olmsted designed the park, including rocky paths, gardens, and a grassy knoll with a view of the cascade. The manicured landscape satisfied progressive calls to return the site to its former, pristine state. When the Niagara Reservation was officially opened, in 1885, one newspaper rejoiced: "The spirit of the wilderness has come back to Niagara."[35] For many tourists and preservationists, walking the winding paths and sitting on the lawns directly around the falls evoked the sense of wildness and beauty they imagined that previous visitors had beheld.

The project of protecting Niagara as a site of the natural sublime and pristine, protected nature would be derailed the very next year, in 1886, when Thomas Evershed was granted rights to harness the falls for industrial power. Within ten years of the legal protection, the opening of the Niagara Falls Power plant sparked an industrial revival. Aluminum and chemical companies such as the Pittsburgh Reduction Company, American Cyanamid, Union Carbide, International Acheson Graphite, and the Carborundum Company built factories in Niagara to take advantage of the cheap, abundant electric power coming from Power House No. 1.[36] Industrial development also drew a workforce, and Niagara's population more than doubled between 1890 and 1900. It saw another burst in the following decade, and in the twenty years between 1890 and 1910, the population of Niagara Falls grew from nine thousand to thirty thousand.[37]

The first powerhouse, completed in 1895, tapped just a fraction of the falls' force to generate an unprecedented 50,000 horsepower. To put this achievement in perspective, consider that in 1893 the various oil- and coal-burning plants providing electricity for the Chicago World's Fair—8,000 arc lamps, 130,000 incandescent lights, and hundreds of trolleys, amusement rides, machines, and engines across six hundred acres—had a total capacity of approximately 24,000 horsepower.[38] By 1901 the Niagara Falls Power Company was the largest electric transmission company in the world. A second powerhouse was built that doubled the output to 100,000 horsepower, which represented 20 percent of the total electric power generated in the United States.[39] One reporter suggested that "cheap Niagara Power" was attracting a flood of "industrial concerns to the Niagara frontier."[40] The pace of development was astonishing. Some predicted that Niagara would soon generate 1 million horsepower of cheap, clean energy.

The great conundrum was how to transmit power into an area called the "Niagara frontier." The Niagara frontier is the recognized name of the region around Niagara Falls that comprises the four counties south of Lake Ontario and Lake Erie and along the western edge of New York State. In 1890 work began on the horseshoe-shaped 1.5-mile tunnel to divert the Niagara River through a series of canals, locks, and chutes and into a 178-foot-deep wheel pit. This dangerous work claimed the lives of twenty-eight men.[41] Meanwhile, the International Niagara Commission, headed by Lord Kelvin, considered twenty different proposals for transmitting the harnessed power from Niagara to Buffalo. Only six proposals included electricity. Ropes and pulleys, gigantic steel drive shafts, and hydraulic tubes that would pump water or compressed air between the terminals in Niagara and Buffalo seemed more plausible than electric wires. In 1891 the Niagara Commission awarded Eben Hill of Norwalk, Connecticut, a $1,000 prize for his proposal to send compressed air through pneumatic tubes.[42] Sending the power through air tubes would have required transformer-like machines at either terminus to convert the energy of the spinning turbines into pressurized air and, in Buffalo, to release the air and use it to generate electricity for lighting and other devices. George Westinghouse, who would eventually win the bid to build the electrical apparatus for the Niagara Falls Power Company, initially believed hydraulic air pressure was the most efficient means of long-distance power transmission at Niagara Falls.[43] Earlier in his career Westinghouse had invented a system that moved natural gas through narrow pipes at high pressure and connected to larger-diameter pipes that helped decrease the pressure, so the gas could be distributed to customers. Securing Tesla's AC patents in 1888 and then facilitating the installation of

350 alternating current generators helped to change his mind.[44] Edward Dean Adams later described Tesla's claim about sending alternating current across long-distance power lines as "a prophecy rather than a completely demonstrated reality."[45] The commission began to seriously consider Tesla's claims after the success of the Lauffen-Frankfurt line. In 1891, using an AC system that may have infringed upon Tesla's patents, German engineers and electricians successfully transmitted 190 horsepower of current across 112 miles with 74.5 percent efficiency.[46] With this new breakthrough the Niagara Falls Power Company solicited a new set of proposals for alternating current generators and high-voltage transmission lines. The requested bids came from the two largest manufacturers of electrical equipment in the United States, Westinghouse Company and General Electric.

The bidding competition turned criminal. In Pittsburgh three Westinghouse employees were arrested on charges of stealing blueprints and selling them to General Electric.[47] Nevertheless, on May 5, 1893, a few months before the opening of the Chicago World's Fair, Westinghouse earned the contract to build the power equipment. Westinghouse's team began work on the first three, yurt-shaped, 5,000 horsepower AC generators and auxiliary powerhouse equipment. General Electric, who was at that time developing AC transformers for a plant at Mill Creek in California, won the contract to design the transformers, the substations, and the 11,000-volt transmission lines extending approximately twenty-six miles to Buffalo. In 1895 the company subcontracted the work of installing the poles and wires to the White-Crosby Company of New York (fig. 9).[48]

The attempt to use cables to carry the electric power was viewed as a great experiment. Adams later explained: "When one visualizes the far-reaching importance of that decision [to use electric transmission with alternating current] and the disaster which would have followed from failure, there comes an appreciation of the vision, the imagination, the judgment, the daring and the courage . . . It was an adventure into the unknown."[49] The plan, according to Frank K. Hawley, director of the Cataract General Electric Company, was to bring 20,000 horsepower into Buffalo, then build lines east alongside the Falls Branch of the New York Central Railroad to reach Rochester. Technical articles in popular outlets such as "The Niagara Falls

9. Niagara Falls Power Company field wagon, ca. 1898. Kenneth M. Swezey Papers, Archives Center, National Museum of American History, Smithsonian Institution, Washington DC.

Electric Line" debated the extent to which electric power could be effectively transmitted and at what distance.[50] Edwin J. Houston and A. E. Kennelly claimed that if 150,000 horsepower could be sent from Niagara to Albany across power lines, the cost to the consumer would be much less than building local power plants that utilized coal and steam. As voltages and transmission distances increased, this hypothesis proved correct. Niagara, in this sense, was the model for the future grid: centralized power plants sited in rural areas shipped their electric power to urban sites of consumption.

When the Niagara-Buffalo line was energized, pundits wondered how far Niagara's power might reach.[51] *Cassier's Magazine* predicted, "We may at least imagine and admire a bow of brilliant promise—an arc of electrical energy stretching from the Metropolis of the Atlantic [New York City] to the Metropolis of Lake Michigan [Chicago], whose waters, swelling the mighty flood that stirs Niagara, may then be called upon to drive 'The roaring loom of time itself.'"[52] The reference is to Thomas Carlyle's translation of the Earth-Spirit's speech in Goethe's *Faust*: "'Tis thus at the roaring Loom of Time I ply, / And weave for God the Garment thou seest Him by." With this inflection the Niagara Power plant may have reminded visitors of a loom with various threads streaming from one rack. Elsewhere, and more commonly, the power plant was viewed in terms of harnessing, or constraining, a wild beast. Consider these headlines: "Niagara in Harness" (*Cosmopolitan*, 1894); "Niagara Put in Harness" (*New York Times*, July 7, 1895); "Niagara in Chains" (*Review of Reviews*, August 1895); "Niagara Is Finally Harnessed" (*New York Times*, August 27, 1895); and "The Harnessing of Niagara" (*Blackwood's Magazine*, September 1895).[53] If the great beast of Niagara Falls had been chained, tamed, or harnessed by pioneers, then it seems fitting to think of the spinning steels caps of the 5,000-horsepower generators as something like the bit in the mouth of a mythical steed; the transmission lines carrying the alternating current into the landscape could symbolize the reins galloping in rhythm across the horizon.

After Niagara began transmitting power to the United States and Canada, it was touted for its natural *and* technological wonders. Niagara was a multisensory experience, accessed via sights such as the massive deluge, floodlights, and turbines; sounds such as the slapping water, the "Cave of Winds," and the humming machinery; and the effect of falling, dissolving, or being swept away. When H. G. Wells toured the falls and then visited the Niagara Power plant in 1906, he reported that the 5,000-horsepower dynamos and the spinning turbines "impressed me far more profoundly than the Cave of Winds; are, indeed, to my mind, greater and more beautiful than that accidental eddying of air beside a downpour. They are will made visible, thought translated into easy and commanding things. They are clean, noiseless, and starkly powerful."[54] The power of nature represented by the falls competed with the man-made power drawn from it.

The cables extending from the dynamos and then raised onto the conductors sitting atop sixty-foot shaved cedar poles may have had a revolutionary function, but they looked considerably more benign than the spinning turbines. The superintendent for the Niagara-Buffalo line said its design followed "recognized practices of the period for telegraph lines—the only precedent there was to follow."[55] The ubiquitous form of the Niagara-Buffalo line did not belie the fact that these lines transmitted electric power (see fig. 5). The unique, less visible function gives the scene importance; the horses and wagon parked between the poles seem quaint. Nevertheless, at the time, the sight of these lines promised change. Wells thought transmitting power into the area around Niagara would attract visitors to gaudy tourist sites. "All the power that throbs in the copper cables," Well's worried, might be directed to amusement rides, metal trinkets, and neon tourist signs. Anyone who has visited Niagara Falls could see that Wells was at least partially correct, but the lines also linked to other awe-inspiring gadgets, and their ability to create action at a distance also evoked feelings of the sublime.

In May 1896 Western Union used its telegraph lines to carry current from the Niagara Falls plant over more than four hundred miles to the National Electrical Exposition in New York City. The amperage carried by the telegraph wires was not strong, but it was sufficient enough to power some lights inside the exposition as well as a water pump attached to a diorama of Niagara Falls. American Telephone and Telegraph made another long-distance connection: a receiver was positioned at the base of the falls and connected to a speaker in the city. A team of engineers, including Tesla, was assigned to build the necessary circuits, transformers, and switches. The Niagara-to–New York City connection represented a breakthrough in terms of distance: the power of Niagara Falls could literally and figuratively be transmitted to New York.[56]

The feats of transmitting messages and power across great distances were often disassociated from the lines in the landscape. The fact that telegraph lines could be refitted to transmit power visually coupled the two systems and may have led the sight of power lines to be viewed as rather commonplace. At the Electrical Exhibition, however, visitors could also see "a handsome little case of historical wires." The wires in this display included sections used for Morse's first telegraph, the first transatlantic cable, Bell's first phone call, Edison's first central station in Manhattan, Sprague's first trolley, and of course, the first power line to carry electricity

from the Niagara Falls power plant to Buffalo. These scientific and technological displays reminded visitors of the wires required for the electrical
revolution in their midst.[57]

Considering how close the Niagara Commission came to transferring
the power from Niagara to Buffalo through a series of pneumatic tubes,
it seems unlikely that its members could have predicted that in less than
a hundred years transmission lines would join the railroad as one of the
most ubiquitous symbols of power upon the American landscape. This
groundbreaking power plant and electrical grid charged the national psyche
and the way Americans viewed overhead infrastructure.

Wired Wilderness

Before the bows of brilliant promise crossed the "electrical mecca" of the
Niagara frontier, a telegraph forest grew out of control through New York
City, four hundred miles to the southeast. Recall that in 1872 the Brooklyn-
based artist John Gast had visualized "American Progress" as the mythical, angelic spread of lacy telegraph lines into the West (see fig. 3). In his
hometown his fellow citizens may have scoffed at such visions, as their daily
experiences with telegraph lines painted a different picture. New York was
covered in wiry webs—and they posed grave dangers. In 1876, for instance,
a rotten telegraph pole on the corner of Suffolk and Grand fell over and
crushed an Irish immigrant named Ann McGuire.[58] The fatal accident
revealed the unregulated business of wiring New York. Representatives from
Western Union and the Manhattan Telegraph denied responsibility. The pole
that killed McGuire did not appear in maps of their existing systems, which
implied that a smaller, now defunct telegraph company may have put the
pole in place. At least eleven telegraph and telephone companies operated
in New York in the 1870s and 1880s, each with its own sets of lines. Many
routes were redundant, and if a company failed, it often neglected to remove
its lines. Representatives from Police Headquarters, the Board of Health,
and the Department of Public Works testified that they were unaware of
this specific pole and its potential danger. Their claims were contradicted
by an officer who told the court he had seen the pole and, believing it a
threat to public safety, had sent a report to his superior at headquarters on

July 17, almost two weeks before the pole crushed McGuire. The jury found that the "parties owning the pole are censurable for allowing a pole in that condition to stand." An "owning party" could not, however, be identified.[59]

Alongside coverage of the trial, the *New York Times* published a ripping satire titled "The Telegraph Forest." Any foreigner, the op-ed observed, knows the United States has "vast tracts of woodland" that "await the axe of the pioneer." However, the European who landed in New York City would be surprised to find that its forest has not yet been cleared: "All over the city the towering telegraph poles and wiry foliage show that the work of reclaiming the island from its pristine barbarism is far from being complete." Unlike other American forests, the telegraph forest offered no shade, its trees bore no fruit, and it was inhabited by "savages" with "sharp spurs implanted on their feet" and "armed with hatchets and heavy glass cylinders." The satire was aimed at both Western Union and the New York City officials, sarcastically saying the company would never "invest money in telegraph trees" when its wires could be buried safely underground; the city officials would never allow a corporation to freely occupy acres of public lands in order to "plant ugly and dangerous poles." Although telegraph poles were "ugly enough" to have been designed by a number of leading American architects, the piece mocked, the city's telegraph forest "must be the work of nature." The course of action was clear: "Let us then, rouse up our energies and clear the poles away" so as to "challenge the admiration instead of the wondering scorn of the intelligent foreigner."[60] Of course, neither McGuire's death nor the satirical "Telegraph Forest" editorial effected immediate change. The wire nuisance remained.

Shortly thereafter, wires for street lighting began to accompany the telegraph wires in their creep through the city. In 1880 the Brush Electric Company opened its first power station for public service at 133 West Twenty-Fifth Street and used overhead lines to carry 2,000 volts to streetlights along Broadway. Edison's Pearl Street Station, the nation's first central service power plant, opened in 1882. By 1884 Pearl Street delivered electricity to over five hundred customers and ten thousand lamps in the area around Wall Street. Edison foresaw the tide turning against overhead infrastructure, and he insisted that the lines for his grid be buried underground.

The cost of burying lines underground was severe and the debts Edison incurred to put the lines underground threatened to bankrupt the Pearl Street project in its first years of operation. Although having underground lines appeased aesthetic concerns, it also posed a threat. Leaked current charged the bricks and stones in streets downtown and were transmitted through the metal horseshoes nailed to horses' hooves, causing the animals to prance in pain.[61]

In 1884 John Trowbridge, a physics professor at Harvard, suggested that the public was willing to accept overhead wires considering that "if electric-light companies were compelled to put their wire underground," then lighting would not be economically viable. He also warned, the public was "only beginning to realize the dangers of the present method of running electric-light wires." A heavy storm that brought lighting wires in contact with telegraph or telephone lines would produce "disastrous conflagrations."[62] Edison explained that neither underground nor overhead high-voltage wires would "remedy the existing evil"; rather, "the only way in which safety can be secured is to restrict electrical pressures."[63] Edison argued that his direct current system, buried underground and limited to 600 to 700 volts, provided the solution. Nevertheless, numerous competing electric lighting companies strung wires carrying different voltages and currents underground and overhead through New York.[64]

In June 1884, in an attempt to clear the telegraph forest and discourage construction of more aboveground electric infrastructure, state senator James Daly went to Albany and won approval for a law that required telegraph, telephone, and electric light companies to bury their lines in cities of more than 500,000 inhabitants. Despite the law, the wiry babel continued to spread. Less than a month after the Daly bill was passed, the New York and New Jersey Telephone Company began to erect poles and string new overhead wires on Fulton Street in Brooklyn. The poles obstructed and in some instances required workers to tear down shopkeepers' awnings. Outraged Fulton Street business owners confronted the workers, and a heated argument ensued. A reporter at the scene said that amid the scuffle, a man who "looked not unlike a retired prize fighter" barged into the crowd and identified himself as "Prescott L. Watson, Superintendent of the Fire Alarm Telegraph." Watson told the shopkeepers that this line had been authorized by the New York City Fire Department and that the Daly law specifically allowed exceptions for overhead lines that helped the Fire Department keep its alarm system operational. Construction on Fulton Street moved forward. The reporter was skeptical of Watson's story and implied to readers that the "Underground Wire Bill" had merely resulted in telegraph and telephone companies forging a "friendly alliance" with "some of the officers of the Fire Department."[65]

Other New Yorkers also brought lawsuits against the telegraph and telephone companies whose wires and poles obstructed their property. In 1884 Judge Van Brunt's response to the case of H. Clausen and Sons' Brewing Company against the Baltimore and Ohio Telegraph Company produced the "sweeping decision" that the poles and wires in New York must "not do injury whatever to the land of the plaintiff in that it does not obstruct the light, air, or access to any portion of the building now erected upon it." Judge Van Brunt added, "Large cities should be freed from the nuisance of having their streets encumbered and disfigured by numerous poles crowded with wires and cables."[66]

The owners of the poles and wires seemed unmoved. In 1885 Daly introduced a supplementary measure to appoint "wire commissioners" to decide which existing lines in Manhattan and Brooklyn must be buried, to produce a reasonable timeline for laying them underground, and to determine how to deal with infractions. Only by burying the lines, Daly argued, could New York have "some chance of becoming a beautiful city as well as a commercial metropolis."[67] A series of legal maneuvers ensued, and both state and city governments took actions "to remedy an evil known to exist in defiance of the will of the people."[68] Western Union, the principal owner of overhead infrastructure in New York City, generally ignored the court's demands. As Jay Gould and his legal team argued, the Fire Department relied on some of Western Union's wires; therefore, the Daly law did not apply.

Ironically, the wires meant to help sound the fire alarms obstructed efforts to fight fires. The New York Board of Electrical Control reported that wires strung across various telegraph and telephone poles ranging in height from forty and ninety feet formed "a complete network, rendering the efficient use of the hooks and ladders and life saving apparatus of the fire department almost impossible."[69] Telephone historian Herbert Newton Casson also noted the challenge of building telephone networks in New York: "Wires had swollen from hundreds to thousands ... Some streets ... had become black with wires."[70] As a rule, during the 1880s telegraph, telephone, and utility companies strove to provide reliable service to attract customers while snubbing the public's desire for less visible, less dangerous infrastructure.

The turning point in this "Battle of the Wire" came in the wake of the deadly Blizzard of 1888 and paralleled the "War of the Currents" between Edison's DC system and Tesla's AC system. In mid-March 1888 twenty-one inches of snow, freezing temperatures, strong winds, and drifts ten to fifteen feet tall buried the Eastern Seaboard. The damage to the electrical infrastructure in New York was severe: hundreds of utility poles came crashing down under the weight of the snow and ice. Telegraph, telephone, fire alarm, and wires for lighting were gnarled into frozen tangles. New York lost communication with Boston, Philadelphia, and Chicago for weeks. Without telegraph contact most cities could only report local news, and communities felt acutely isolated from their neighbors, something they had not felt so sharply before they came to rely on the telegraph.

The underwater transatlantic cables between New York and Europe, however, "suffered no whit of interference or interruption from the storm." Irked by the fact that the news from Europe continued while nearby cities in need of information and supplies were effectively cut off from their neighbors, the *New York Times* reminded readers that American telegraph and telephone networks should, and could, be easily buried underground. In fact, the law had demanded the wires go underground. The paper explained that "the taxpayers of New-York desire ... the poles and wires now standing and stretched along the lines of completed subways removed at once."[71]

Despite the ready availability of conduits along the routes being excavated for subway tunnels, telephone and telegraph companies still "systematically violated" the Daly law by stringing wires at night or without permits.[72] On May 12, as workers continued repairing the region's overhead lines brought down by the storm, the *New York Times* ran a blunt headline: "Bury the

Wires." Undergrounding into the subway conduits was slow but steady, and by the summer of 1888, five hundred miles of wire and 217 telegraph poles had been removed from the streets of New York.

Even as the problem with telegraph and telephone wires seemed to come under control, however, transmission and distribution networks that would serve a new catalog of lighting, factories, and home appliances arose to threaten the skies. On May 16, 1888, Tesla introduced his alternating current system at a meeting of the American Institute of Electrical Engineers at Columbia University (then situated at Forty-Ninth Street and Park Avenue). With step-by-step diagrams Tesla proved that his alternating current system afforded "peculiar advantages, particularly in the way of motors, which . . . will at once establish the superior adaptability of these currents to the transmission of power."[73] Tesla's patented system, purchased and developed by George Westinghouse, foreshadowed the proliferation of electrical grids for lighting and power.

While many scientists and engineers accepted the superiority of Tesla's system, Edison had already borne the brunt of building, and undergrounding, his direct current grid in Manhattan. Edison infamously attempted to convince the public that alternating current, whether above or below ground, was unsafe. A perfect opportunity presented itself to Edison and his team a few weeks after Tesla's presentation. In June 1888 the New York legislature selected electrocution as the state's official method of capital punishment. On July 31, 1888, three employees of Edison Electric Company used alternating current, Tesla's competing system of power, to electrocute dogs and other animals before a group of journalists and engineers in the same room where Tesla had lectured weeks earlier. The horrified crowd watched as one seventy-six-pound dog received "successive jolts of 300, 400, 500, 700 volts of direct current" before being killed by 500 volts of alternating current. If AC was this effective at killing animals, what kinds of risk would AC pose if it were strung through the streets or brought into homes? The public electrocutions effectively spread doubts about the Tesla-Westinghouse model and other alternative systems of electrification. By the end of the following summer a committee, swayed by Edison, decided that the first electric chair would carry alternating current. Even in 1902, long after AC was accepted as safe and the standard means for transmission and distribution, Edison and his film crew recorded the alternating current electrocution of Topsy, a rogue circus elephant in Coney Island.[74]

The Blizzard of '88, as already noted, revealed the vulnerability of overhead wire networks and the public's dependence on railways, telegraphs, and telephones. Meanwhile, staged electrocutions and a series of electrical accidents created a culture of fear that Joseph P. Sullivan calls "overhead wire panic."[75] Between May 1887 and September 1889 seventeen New Yorkers were killed by electric currents, many of them from faulty wiring in the streets. The most gruesome spectacle occurred in October 1889, when John E. H. Feeks was tangled in the wires stemming from a fourteen-crossarm pole on Chambers Street near City Hall. The smoking, charred corpse hung above a horrified crowd for almost an hour until it could be cut down.

A few months later Mayor Hugh J. Grant—who was just thirty-one years old when elected in 1889—ordered that "dead telegraph poles" be chopped down and their wires removed. Grant marked specific sections of Manhattan for demolition: "Third Avenue, Railroad Avenue and One Hundred and Thirty-Ninth, Wooster, West Seventy-Third, Park, Centre, Leonard, Bayard, Hester, Broome, Mercer, and Spring Streets, Bowery, Fourth Avenue, and Park Place."[76] The public supported the mayor's clearing of the telegraph forest, and crowds gathered to cheer on workers as they chopped down the poles (figs. 10–11). In his 1947 memoir, *Manhattan Kaleidoscope*, Frank Weitenkampf recalls the New York of his childhood with "the sky almost obscured by criss-crossing wires" and the "special treat" of watching lumbermen who had traveled to the city to cut them down. He recalls being "thrilled by the steady, sure sweep of the long-handled axes which they wielded."[77] Historian Mary Cable adds that as "the poles began to fall like forest oaks," "crowds followed the workmen and cheered and cried 'Timber!' as the poles hit the ground . . . At last the city could see the sky again."[78] At the end of the year Public Works commissioner Thomas Gilroy reported that 731 telegraph poles and 1,194 miles of telegraph wire had been removed from the streets of New York.[79]

As the telegraph forest was cut and cleared, new wires for lighting seemed to sprout up in its place. A feature piece on "Electric Lighting in New York," published in *Harper's Weekly* in 1889, shows the lines for electric power extending from the Brush Electric Light Company at 210 Elizabeth Street. The article suggests the difficulty of comprehending the function of such a small and simple-looking line: "The observer in the street sees only the silent wire which if it is the size of a lead-pencil he considers rather large. If this wire is supplying its usual quota of fifty arc lights, it is carrying from house to house about fifty-horse-power, which, being entirely invisible, is likely to lack appreciation." The observer would have difficulty imagining how "silent electric wires in the street" could carry as much power as the "older and purely mechanical means of transmission" that required wheels, pulleys, belts, and pipes.[80] Again, the services provided by, and technologies attached to, the wires seemed astonishing, but the silent and visually unremarkable form meant that most viewers would likely not understand how the lines functioned.

A few years later, when boosters promoted the idea of an electric streetcar system for New York City, the aesthetic complaints resurfaced. *Harper's* reminded readers that "great efforts have been made in New York and other cities to compel the burying beneath the ground of all electric-light and other wires through which currents of electricity are passed." The new lines for the trolley would be another "sad disfigurement of the public streets," and "every objection that was good and valid against the electric-light, the telegraph, and telephone wires may also be urged against these wires for the trolley."[81] It seems fortuitous that a compromise for undergrounding the telegraph and telephone lines had been reached in the early 1890s, as it helped set a precedent that was demanded of subsequent electric utilities and trolley companies in New York City. And the wave of wires kept coming. By 1893 twenty distinct light and telegraph companies serviced Manhattan, each one using "independent sets of wires carrying varying voltages (if direct current systems) or operating at distinct rates of oscillation (if alternating current systems)."[82] Of course, in the boroughs where the trains were not buried—most notably the Bronx, Queens, and Staten Island—overhead telegraph, telephone, and distribution lines continued to spread overhead through public spaces and to obscure views.

In most American cities electrification still meant upright poles and

FIG. 25.—DISORDERLY WIRES ON LOWER BROADWAY ABOUT TO BE CUT DOWN.

10. This illustration shows the "disorderly wires" about to be cut down near the intersection of Broadway and John Street in lower Manhattan. Schuyler S. Wheeler "Electric Lighting in New York," *Harper's Weekly*, July 27, 1889, 601.

BROADWAY AND JOHN STREET, NEW YORK, AS IT APPEARS
WITHOUT OVERHEAD WIRES

11. A photograph from a similar perspective of the intersection of Broadway and
John Street "as it appears without overhead wires." Herbert Casson, *History of Telephone* (Chicago: A. C. McClurg & Co., 1910), 133.

looping wires, and communities were routinely disgusted by the visual impacts. Overhead wires signaled spaces that were unkempt and even unclean. In the early twentieth century urban planner Charles Mumford Robinson declared that cities such as Raleigh, North Carolina, suffered from "wire evil" and warned that Binghamton, New York, "could not, even with its quite metropolitan sky scrapers, seem more than a country town, as long as the pole and wire evil continued in so aggravated a form."[83] Overhead wires signaled a kind of moral turpitude.

Meanwhile, within Manhattan well-lit and uncluttered public spaces offered pedestrians new vistas and orientations. Without the obscuring wires, Manhattan could be promoted as a city of long avenues, river promenades, rising skyscrapers, and streetlights. As the century turned, streetlights wound through Central Park and began to concentrate in places such as Times Square. The wires, still pervasive and increasingly critical, were hidden from view. Burying the lines may have protected them from bad weather and aesthetic blight, but being out of sight did not always put them out of mind. In fact, the inability to see the wires made them seem more vulnerable to manipulation by certain corporations and individual "wiretappers."

The Wire Tappers and Phantom Wires

Electric technologies such as the telegraph and the telephone made significant appearances in "electric" literature of the late nineteenth and early twentieth centuries. Numerous scholars have examined electric technologies in Ella Thayer's *Wired Love: A Romance of Dots and Dashes* (1880), Mark Twain's *Connecticut Yankee in King Arthur's Court* (1889), Henry James's *In the Cage* (1898), and Frank Norris's *Octopus: A Story of California* (1901).[84] Arthur Stringer's crime and suspense novels, *The Wire Tappers* (1906) and its sequel, *Phantom Wires* (1907), have received little scholarly attention, but they reflect three of the tropes and patterns scholars have identified in other electric literature.

First, like *Wired Love* and *In the Cage*, the telegraphic communication both facilitates and complicates romantic exchanges. The protagonists, American Jim Durkin and his British partner, Frances Candler, use telegraphy to send sappy messages, and the technology inspires sentimental connections between electricity and emotion. At one point, bending over a receiver, Durkin asks: "What is electricity? . . . We live and work and make life tenser with it, and do wonders with it, but, after all, who knows what it is?" to which his partner replies, "And what is love? . . . We live and die for it, we see it work its terrible wonders; but who can ever tell what it is?"[85] Electricity keeps the lovebirds connected, even if the means and materials of those connections remain uncertain. Second, the couple fluently codes and decodes Morse messages, sometimes using impromptu tools and methods such as blinking eyes or chattering teeth to "tap" warnings or clues in the presence of unsuspecting bystanders. Using this technique, Jim and Frances pretend to not know each other and then carry on conversations in front of their adversaries. Such acts blur the boundaries between private and public communication.

Similar to *The Octopus*, the novels also reflect populist fears that criminals and corporations manipulated the flow of information over the wires. Durkin imagines "the four great Circuits, Eastern, Southern, Western, and Pacific slope, of the huge and complicated and mysteriously half-hidden gambling machinery close beside each great centre of American population."[86] The sheer pervasiveness of the gambling system (and the stock and commodity exchanges) are only "half-hidden"; therefore, the wires are exposed to criminals, like Durkin and Frances, who can tap into the system and alter it to their advantage.

What distinguishes Stringer's novels from other electric literature is the protagonists' electrical expertise, which allows them to tap, cut, and splice wires. Durkin is an inspiring inventor. He plans to use "Tesla currents" and selenium cylinders to create a "transmitting camera" (or what today would be called a fax machine). Short on research funds and blacklisted as a telegraph operator, he agrees to aid in a gambling scam with a notorious wiretapper named McNutt, whose gang also includes Frances, with whom Durkin quickly falls in love. Durkin and Frances eventually turn against McNutt, who they deem is a greedy, disgusting, and depraved criminal. They continue to tap wires and steal cash, diamonds, and sensitive information, but throughout both novels Stringer portrays them with refinement and detachment. When Durkin climbs onto the roof of a poolroom in Greenwich Village, for example, he removes from his bag "a Bunnell sounder, and then a Wheatstone bridge, of the post office pattern, a coil of KK wire, and pair of lineman's pliers, and a handful or two of other tools." He goes to the "tangle of insulated wires issuing from the roof," then "skillfully relax[es] the metallic cable strands" and "carefully graduate[s] his current and attache[s] his sounder, first to one wire and then to another." Distinct from McNutt and other wiretappers, who are referred to dismissively as "lightning slingers" and "overhead guerrillas," Durkin's wiretapping is "not unlike the difficult and dangerous operation of a surgeon."[87]

If Durkin is the surgeon, Frances is the sophisticated spy. She assumes diverse personas and backstories to pass between networks and social classes. The mobility seems to make her more cynical. She claims, for example, the problem with Americans' risky investments and gambling via the telegraph is not "the spectacular way of your idle rich wasting their money" but that of the culture of risk and big wagers without capital. This detachment allows the lower classes to become infected with the "diseased lust for gain without toil." With no clear means to achieve justice and release the poorer classes from this lust, she steals to undermine the corruption of wealth and power. "The criminal," Frances explains, "lay[s] claim to a distinct economic value, enjoining . . . continual alertness of attention and cleanliness of commercial method." The wiretappers' role, in Frances's view, is to infiltrate networks and then use the stolen information as a kind of bacteria, a "natural and inevitable agent" that will infect the wires and therefore make the system cleaner. Frances even appreciates this amoral disinterest in one of her violent enemies, Keenan. Keenan is a disbarred lawyer, and Frances, pretending to fall in love with him so she and Jim can steal the telegraph company bonds he has stolen from a mob boss, sees Keenan as a "criminal of such apparent largeness of mind and such openness of spirit that his very life of crime . . . take[s] on the dignity of a Nietzsche-like abrogation of all civic and social ties."[88] The wiretappers may operate in dark, secret places, but they are united in their dignified challenges to corporate power.

While wires offered an analogy for the various characters and messages that coalesce in the fictional network, Stringer did not venture far from the headlines when crafting plots for his protagonists. In *The Wire Tappers* the couple prepares their horse racing scam by betting regularly at various poolrooms around New York City. When the time comes, they will make a large wager on a horse that, through their wiretaps, they know has already won at a distant track. They will collect the winnings and disappear. Their plan matches real-life crime of the era. The most popular version of such scams required tapping into lines in order to obtain race results before they had been delivered. The tapper could go into a parlor and "bet" on the winner before the simulcast of the race, which was delayed ten or fifteen minutes, had begun. Another way to cheat the system was by interrupting the circuits and sending false race results to the gambling parlor. An exposé published in the *National Police Gazette* in 1902 said that wiretappers had "systematized the art to such [a] basis that they never are arrested, travel from place to place where poolrooms exist, live in the best hotels . . . and do no business other than tap wires."[89] By the early twentieth century gambling establishments had taken precautions against wiretaps either by prohibiting last-minute wagers or delaying payouts until results could be confirmed through a second line.

Stringer also fictionalizes the insider trading that took place on the commodities market. After running their horse racing scam, Durkin learns what "every respectable broker" already knew: someone is intercepting the monthly cotton reports sent from New Orleans to Washington DC and manipulating the futures market. Durkin taps this "Machiavellian operator's private wires" and discovers that a New York businessman referred to as the Cotton King is preparing to push the price of cotton exorbitantly high, near twenty cents a bushel, only to burst the bubble and ruin his competitors.[90] The Cotton King may have been based on powerful brokers of the era such as J. P. Morgan or Western Union owner Jay Gould,[91] yet the imaginary scene of wiretapping, wild speculation, and nearly instant collapse that produces the novel's climax was based on a different, but no less real, historical fact. On September 29, 1899, a group of wiretappers hacked into the international telegraph line transmitting cotton prices from Liverpool to New Orleans. Within a few hours the price of cotton futures rose so quickly that it caused "the wildest panic ever witnessed on the floor of the New Orleans Cotton Exchange."[92] In the years following the historical events of 1899, the narrative setup of a prearranged market panic appeared in fiction and film. The public seemed fixated on how markets linked by complex technologies, such as telegraphy, were vulnerable to white-collar criminals who could control networks, set prices, and ruin innocent investors.[93]

Concerns that wire networks would encourage financial corruption are as old as the telegraph itself. Wiretappers and shady traders worked to cheat the market. By the end of the nineteenth century, as corporate power took a firmer grip on the American economy, many Americans were convinced that stock and commodities exchanges, telegraph companies, major media corporations, and wire hackers worked in cahoots to lie, cheat, and steal for the sake of power and profit.[94] (Of course, such narratives have been recycled and extended. The 2008 global financial crisis, for instance, has been attributed to increasingly interconnected markets, the advent of high-frequency online trading systems, and nefarious investment strategies.)

Jim and Frances send and receive, code and decode, across wires; they also tap *into* the wires. This sense of being embedded in their wired environment separates these works from later wire thrillers, such as Frank Packard's novel *The Wire Devils* (1918). The dubious intercepting of secret messages and last-minute escapes are bookended by soliloquies, in which Jim and Frank (as he calls his partner) make explicit the connections between electric transmissions and consciousness. *Phantom Wires* opens with the couple having safely escaped McNutt in New York and now residing in a European resort town. Durkin tells Frank, "Our wires are down, for a little while anyway." She disagrees, telling him: "There are always some sort of ghostly wires connecting us with one another, holding us in touch with what we have been and done, with our past, and with our ancestors, with all our forsaken sins and misdoings . . . There are always sounds and hints, little broken messages and whispers creeping in to us along those hidden circuits. We call them Institutions, and sometimes we speak of them as Character, and sometimes as Heredity, and weakness of will—but they are there, just the same!"[95]

It appears Frances is mixing metaphors—the "ghostly wires" that hold us in touch with the past seem like telegraph wires. Later the "hidden circuits" that send "broken messages and whispers" act like telephone lines. In both cases the passage relates to a kind of phantasmal or unseen network. Such networks, like the "spiritual telegraph" of the mid-nineteenth century, transcend time and space. The significant difference here is that this invisible network has the ability to alter seemingly stable institutions and character traits. The ghostly wires connecting the species are like DNA, linking us with ancestors, controlling our character traits, and tethering humankind past, present, and future in one vast, grid-like genome.

By the end of the *Phantom Wires* Durkin makes an associative leap from the idea of "ghostly and phantasmal wires connecting mind with mind" to the idea that the soul is an "elusive wireless." The connection between the soul and consciousness is like the connection between two wireless devices. He realizes: "Yes, consciousness was like that little glass tube which electricians called a coherer and all his vague impressions and mental gropings were those disorderly minute fragments of nickel and silver which only leaped into continuity and order under the shock and impact of those fleet and foreign electric waves which floated from some sister consciousness aching with its undelivered messages."[96]

This is the final stage in Durkin's conversion from a "wired" to "wireless" mind-set. Previously, Durkin considered consciousness like "a half-articulate key, at the end of an impoverished circuit." The thoughts that one materializes, even "the most artful and finished," are "merely a sort of clumsy Morse."[97] In this metaphor the telegraph models how half-formed and indecipherable ideas appear in consciousness. By the end of the second novel Durkin's view of consciousness has evolved. Rather than being sent and received over a telegraph wire, thoughts are more like the vague, fleet, and foreign waves sent across wireless devices. More specifically, consciousness is like a coherer, a glass vacuum-sealed tube with small metallic fragments that are influenced by the transmission of invisible electric waves. In Durkin's view the brain sends and receives thoughts and ideas to consciousness via Hertzian waves that radiate through space.

The concept of wire*less* consciousness is contingent on a general understanding of wires, the metallic threads that transfer electricity between points in a circuit. In other words, a world filled with physical wires is required to convey metaphors related to wireless consciousness or the ghostly circuits that transmit hidden messages, remember behaviors, and store genetic information.

Tesla first introduced wireless technologies in the 1890s, and by 1907 Guglielmo Marconi's wireless devices had been reproduced and improved. Stringer's novels thus show wired and wireless networks interlacing in public discourse. The telegraph poles and wires that had previously guided and framed the landscape were beginning to disappear. Meanwhile, similar-looking poles and lines leading to home appliances, street lighting, and factory areas were being strung through the landscape. Basing a meta-physical belief or electrical metaphor on a wireless system does not offer the same tangible tenor, concrete association, or visible infrastructure. In the decades after Stringer, modernists took advantage of the possibilities offered by thinking in terms of wireless networks, universal waves, and sites inundated by signals with certain frequencies. The boundaries of these waves and fields are not marked by the physical wire in the landscape but by various tuning dials and ersatz maps showing the places one might hope to pick up a signal.

An Empire without (Visible) Wires

The tangled web of telegraph, telephone, and electric power wires that spread overhead and underground in New York and other parts of the nation had at least two unintended effects. First, the expansive infrastructures were difficult to secure and could be compromised by criminals who cut, spliced, and tapped lines. The channels of communication and power on which businesses and the public increasingly relied could be compromised (and sometimes erased). The second unintended effect of a wired nation was the spread of electrical metaphors through popular culture and everyday language. Like the nineteenth-century representations of the telegraph, lines and wires continued to unite individuals in time and space and allow for relations between individuals and distant groups, cultures, and races. By the turn of the century, however, the meanings and possibilities for electricity had evolved. Electrical devices and lines proliferated through the landscape with greater magnitude and density. As wireless devices came onto the

American scene, they helped to increase the general awareness of electric technology's impacts on the processes of thought and construction of self.

In Manhattan, the central node of the Empire State and increasingly of the world's financial markets, the wire blight was alleviated, but the material infrastructure buried below the ground was still vulnerable. Across the American landscape overhead wires were not only vulnerable to wiretappers; they were a visual nuisance subject to disorder, tension, even rupture. Expanding networks for electric lighting and power exacerbated the wire problem in many American cities. Between 1902 and 1907 the mileage of telegraph and telephone wires in the United States nearly doubled, and of the fifteen million miles of wire operating in 1907, almost 80 percent of these wires were strung from poles, rooftops, or other overhead structures.[98] Telegraph, telephone, and power companies built their networks with similar-looking wires and sometimes used the same increasingly overburdened poles. A few municipal authorities began to require telegraph, telephone, and utility companies to bury their lines, but the lines continued to multiply, and to most viewers, the lines remained indistinguishable.

3

California's Wood Poles, Steel Towers, and Modernist Pylons, 1907-1972

The wood pole lines, lattice steel towers, and modernist pylons that have crossed California reflect a unique confluence of technological innovation and environmentalism. According to Donald Worster, the freedom evoked by western landscapes and the comfort offered by new technologies and infrastructures made Californians "especially receptive to the vision of a technologically dominated environment."[1] The electricity carried by overhead wires supported the broader vision of the California dream—a dream of warmth, comfort, beauty, and wealth. The lines undergirding that dream represented subjugated landscapes.

Geographic, political, and economic forces distinguished California's process of electrification. During the later stages of the gold rush and then the Civil War and Reconstruction, the Golden State seemed to cling to the rest of the nation by feeble threads. At the turn of the twentieth century expansive transmission networks connected California's burgeoning systems of lights, factories, and railways. California's telegraph lines and its electric power lines sent distinct cultural messages. The softened messages appeared in newspapers and trade journals, black-and-white films, and the Dreyfuss designs, which represent the single significant effort to build "aesthetic" transmission lines in the United States. Advertisements released by Southern California Edison (sce), one of the state's largest utilities, also promoted visions of California as a place of sublime scenery and advanced technology. Electric lines, as material and metaphor, facilitated the extension of a so-called second nature across California. Over time, as the lines multiplied, the public pushed back against the lines encroaching upon the natural environment, and California emerged as a leader in the long-distance transmission of power as well as in public opposition to overhead lines across the landscape.

A cultural history of California's networks can be culled from select, sometimes unintended engagements with overhead lines, including those instances that suggest the lines' aesthetic impacts. For instance, telegraphy, as a quick, exciting medium and an alternating, dialectic mode, has been tied to the advances in narrative structure and editing in the early days of American cinema. The electric lines (some of which carried power) that appear in the background of almost every outdoor shot of D. W. Griffith's *Lonedale Operator* (1911) and *The Girl and Her Trust* (1912) also reflect the less well-known intersections between the western genre, Hollywood, and the tension lurking beneath a wired society. Meanwhile, California's remarkable history of power transmissions provides crucial context for understanding reactions to these wiry artifacts.

California's long-distance transmission boom began in a narrow canyon of the San Bernardino Mountains in 1892. As Nikola Tesla and other engineers promoted new alternating current systems for lighting and power transmission in the East, Almerian Decker, a self-taught electrical engineer who moved to Southern California to ease the pains of tuberculosis, put AC to distance tests in the West. For his first transmission project, Decker worked with Dr. Cyrus Baldwin, president of Pomona College. Baldwin had organized the San Antonio Light and Power Company to harness the falling waters near the "Hogsback" ridge of the San Antonio Creek. The creek flowed into the citrus orchards surrounding Pomona. Baldwin raised funds, secured permits, and reassured farmers in the valley that their irrigation waters would not be disrupted by the attempt to dam the creek. Decker designed the transmission system, calculated the potential electrical output to be delivered to nearby cities, and tried to convince an eastern manufacturer to build specialized equipment for his experimental power line. The final task proved difficult. As the Niagara Commission considered ropes and air tubes for the transmission of hydropower, Decker's plan to harness a creek and transmit AC current at 10,000 volts across fourteen miles of rocky terrain seemed impractical. Westinghouse Company controlled Tesla's AC patents and built AC central stations to serve customers within a radius of up to five miles, but it initially refused Decker's request to build the necessary AC generators and transformers to complete this longer distance transmission line.

Baldwin traveled to the East Coast to solicit support. William Stanley, who had developed an AC transformer for Westinghouse in 1885, examined Decker's plans and "promised to manufacture the machinery himself" if other companies would not.[2] Stanley's approval helped to convince Westinghouse Company to accept the contract. Decker oversaw the installation of the 120-kilowatt generators and 6.5-kilowatt oil-filled transformers—the first in the United States to step up and step down voltage for transmission. In December, a month after the fourteen-mile, 10-kilovolt single-phase line of number 7 copper wire carried power from the San Antonio plant to Pomona, another 10-kilovolt line stretched twenty-nine miles to San Bernardino. As these lines neared completion, Decker signed a contract with the newly formed General Electric Company to attempt three-phase AC transmission in a nearby canyon named Mill Creek.

In the summer of 1893, as Chicago was abuzz with the World's Fair, the machinery for the world's first commercial, three-phase power system left Schenectady, New York, bound for Redlands, California. The 2,400-volt potential of these 250-kilowatt generators would be "stepped up," similar to the procedure conducted at the San Antonio installation. Rather than transmission at a single phase, however, the three-phase transmission from Mill Creek No. 1 would be transmitted across three parallel cables. Each of the three cables would ideally transmit at the same voltage and frequency: in this case 24 kilovolts at 50 hertz.[3] Tragically, Decker would not live to see his three-phase system completed or the same transmission scheme repeated across the world (today transmission lines operate at 60 hertz, but the live wires often appear in sets of three). Decker's health worsened that summer, and in his final weeks he was wheelbarrowed more than two and a half hours on a path through the canyon to oversee construction. Just four years older than Tesla, Decker died at the age of forty-one, weeks before Mill Creek Power Plant No. 1 made history as the nation's first (and currently longest-operating) commercial three-phase power plant.

The ensuing spread of California's hydropower systems and power lines did not follow a general pattern of city to country or East to West made by previous infrastructures such as the telegraph and railroad. Nor did they appear in the same bric-a-brac manner as most electrical grids. As David Nye suggests, across the United States the "electrical landscape sprang up in patches."[4] However, in the two decades after Decker's successful demonstrations, thousands of miles of increasingly longer, higher-voltage transmission lines spread like the guy wires from high-altitude California creeks, canyons, and mountain streams to distant metropolitan markets. These point-to-point transmissions also included branch lines that provided service to smaller cities and agricultural towns en route to the cities. Aggressive development of hydropower resources expanded and connected California's manufacturing and agricultural capabilities. By the early twentieth century a relatively robust transmission grid had been stretched across the Golden State and helped transform this relatively inhospitable region into one of the most iconic, diverse, and energy-rich regions on the planet. News of California's electrical engineering feats circulated in newspapers and trade journals; images of its landscapes were projected onto the silver screen.

Telegraph Lines, Early American Cinema, and Isolated Landscapes around Los Angeles

For much of the nineteenth century Los Angeles was a small, rough, nearly lawless cattle town. In the last decades of the century, increased oil production, unchecked land speculation, and the arrival of Southern Pacific's intercontinental railroad line each ignited development. Between 1900 and 1910 the population of Los Angeles tripled, from 102,479 to 319,198.[5] The exponential growth was linked to oil and agriculture as well as to the city's electric-powered train system. In 1904 *Harper's Weekly* said Los Angeles had "the most perfect electric transportation system in America" and explained how the "lines of this electric [railway] system lead through a country made famous by its natural beauty and perfected by development."[6] In the new suburbs electric trains and the lines bringing power to those trains seemingly "perfected" the California landscape. Within the city limits the new power lines made more unsettling impressions.

By 1910 Los Angeles boasted the largest consumption of electrical energy per capita of any city in the United States.[7] Telegraph, telephone, and power lines in downtown Los Angeles had been subject to an undergrounding ordinance passed by the city council in 1897. Still, as in New York and other cities, many areas of Los Angeles were inflicted with what *Electrical World* called the "pole nuisance." In 1907 the Los Angeles city government and local utilities reacted by forming the Joint Pole Committee.

Between 1907 and 1909 the Joint Pole Committee of Los Angeles scanned streets, alleys, and rights-of-way for redundancies. Shorter wooden poles that had stood side by side were removed and replaced by single rows of taller poles with three tiers: one for telegraph, one for telephone, and the highest for low-voltage distribution lines. Over the course of two years utility workers combined wires on over ten thousand poles. The new three-tiered poles, "while never exactly ornamental," could "give service more cheaply and effectively than the usual tangle to be found in the streets."[8] Locals may have looked at the three-tiered poles as signs of civic cooperation, progressive urban design, or the positive absence of a distracting tangle. It is unlikely that anyone viewed them as reminders of the Old West. Nevertheless, a row matching the "three tier" description appears in one of the first and most famous westerns of early cinema.

The film bulletin for *The Lonedale Operator* (1911) describes its setting as "the most isolated spot in the Western country."[9] The outdoor scenes in this seventeen-minute silent classic convey isolation: the grove where the telegraph operator (who signs off "M.D." and is played by Blanche Sweet) flirts with the train engineer, the rustic station where she is attacked, and the sandy landscape through which the engineer races to her rescue. Of course, "Lonedale" was not actually isolated. The set was built in the Biograph Studio at Georgia Street and Pico Boulevard (at the current site of the Staples Center). The outdoor scenes were shot in nearby Inglewood.[10] A still image of the train depot shows three wooden utility poles configured for telegraph, telephone, and low-voltage distribution lines (fig. 12). The film may convey the feeling of a lonely frontier outpost, but parts of the Los Angeles power grid poke into the background.

In 1910 D. W. Griffith, cameraman Billy Blitzer, and a troupe of Biograph

12. Still from *The Lonedale Operator*, 1911. The setting is a remote western town, connected to civilization by train tracks and a telegraph thread, but in this shot viewers can see some of the electric power lines that had begun sweeping through Los Angeles. *Lonedale Operator*, dir. D. W. Griffith, Biograph Studios.

actors, including a seventeen-year-old Mary Pickford, made their first winter sojourn from New York City to the West Coast. Griffith and other filmmakers came to Southern California both to escape the legal threats of Thomas Edison's Motion Picture Patents Company and to take advantage of the natural sunlight, the Mediterranean climate, and the diverse scenery— coastline, mountains, valleys, ranches, orchards, missions, and deserts all within a day's commute. During Griffith's first trip he wrote and directed dozens of one-reel films, including *In Old California*, which is considered to be the first film made in Hollywood.

During subsequent trips to Southern California in 1911, 1912, and 1913, Griffith made hundreds of films, averaging between two and three films a week. In these four years Griffith helped to advance American cinema "from the crude assembly of unrelated shots into a conscious, artistic device."[11] He experimented with new narrative forms, camera angles, and editing techniques. By the time Griffith left Biograph Studios, in 1913, he was the most prolific of American filmmakers; after the release of his first feature-length film, the financially successful and grossly racist *Birth of a Nation* (1915), Griffith would remain one of the country's most controversial. Griffith later boasted he was "the Man Who Invented Hollywood." He was not *the* inventor, but he did make major innovations in how films were conceived, created, and distributed to the public.[12] Griffith laid groundwork and helped push the nexus of the industry from New York to Los Angeles, where he, his crew, and the rest of the Angelinos were surrounded by power lines.

The Lonedale Operator's ninety-eight shots have undergone multiple scholarly analyses.[13] Film scholars have seemingly overlooked the three-tiered utility poles or the apparent anachronism of showing telephone and low-voltage distribution lines connecting to an "isolated" telegraph station. On the one hand, the power lines undermine the films' accuracy and authenticity. It is difficult to believe that Lonedale is the "most isolated spot" if we can see that it has been wired for telephone and power. One the other, the presence of the utility poles in the background reinforces the importance of electric technologies and infrastructures to early American cinema.

Film scholar Paul Young says it would be "hard to imagine the complex story films made in the United States without the telephone and telegraph," in part because the cinema aspired to be like the telegraph, a "thrilling new gadget and a carrier of messages—news, spectacles, stories, emotional and visceral effects."[14] Telegraphy was a spectacle; it was also a primer. Audiences generally understood that the telegraph "annihilated" space and time; filmmakers like Griffith displayed characters using the telegraph to creatively stitch space and time back together. Showing M.D. tapping on a telegraph key to send messages implies that the she is connected in real time to the receiver in the next shot. The subsequent shots may appear one after another on screen, but the viewer can imagine, with the help of the telegraph, that actions are happening simultaneously in distinct locations. Griffith's back-and-forth shots between operators is an early example, possibly the first in American cinema, of what is now called crosscutting, or parallel editing. Tom Gunning suggests, "Griffith edits [*The Lonedale Operator*] as if inspired by the telegraph."[15] Film historian Raymond Bellour says that *The Lonedale Operator* epitomizes the ability of film to alternate between characters, settings, and motifs. Like a series of breaks in electric current or dots and dashes sent back and forth, the shots alternate between He-She, she-operator, and she-robbers-train. Tension builds through "slightly displaced analogies, an accumulation of small differences."[16]

Both *The Lonedale Operator* and *The Girl and Her Trust* follow the same plot. In the opening shots young men working for the railroad vie for the female operators' attentions. The operators resist. They sit at their desks and receive telegrams about large sums of cash set to arrive on the next train. The cash arrives. Before the heroines are left alone, male protagonists offer them guns for protection. The second operator, Grace (Dorothy Bernard), scoffs when the station agent whips his gun near her face and tells her to take it. "Danger? Nothing ever happens here," reads the intertitle. Summarily attacked by pairs of bandits, the operators tap out desperate pleas. M.D. sends the message: "Thieves breaking in Lonedale station. Am alone. Send help quick." Grace sends a shorter and thus a bit more terrifying dispatch: "Help . . . tramps . . . quick."

Telegraph lines transmit their messages from the isolated spots where "nothing ever happens," and the operators' beaus are dispatched to make last-minute rescues, facilitated by the train tracks that connect, as the bulletin suggests, "Lonedale to civilization." The tense conclusion of *The Lonedale Operator* was, according to the studio advertisement, "without doubt the most thrilling picture ever produced." The next year *Moving Picture World* magazine called *The Girl and Her Trust* "without exception, the most thrilling pursuit ever depicted." The thrills are fueled by Griffith's masterful interplay of trains, guns, and the threat of violence combined in relatively quaint, working-class landscapes.[17]

Griffith used inventive camerawork and creative film syntax to present the dominant technologies of movement (railroad) and violence (gun, or faux gun). He also included, possibly intentionally, utility poles in almost every single outdoor shot. Collectively, these shots send another stark message about technology in the western landscape.

13. *The Girl and Her Trust*, 1912. Grace seems to sense she is being watched. After she telegraphs a plea for help, she watches as her attackers cut the telegraph wire emerging from her office. This appears to be the first instance in American cinema when a potential victim hears or sees the line "go dead." *Girl and Her Trust*, dir. D. W. Griffith, Biograph Studios.

Poles and wires play an especially important role in *Girl and Her Trust*. Unlike M.D., who faints after sending her call for help, Grace sends her sos and then watches as one of the attackers cuts the wire outside her window. Trapped, uncertain if her plea has been received, and now unable to contact the outside world, she wrings her hands and mouths, "Dammit!" This wire cutting occurs just before Grace's message is decoded at the next station, and within the back-and-forth shots between stations, Griffith flashes to her beau, Jack, standing in a field of telegraph poles (fig. 13).

The crosscutting between locations adds tension to what seems to be the first "wire cutting" scene in American cinema.[18] One of the other iconic early westerns, *The Great Train Robbery*, also begins with an attack on a telegraph operator. After *The Girl and Her Trust* the trope of the feeble operator and wire cutting to increase danger became a staple of the western genre. Two of the more famous examples include John Ford's *Stagecoach* (1939), which opens with Apaches cutting telegraph lines in preparation for an attack on a wagon train, and Sergio Leone's *Once upon a Time in the West* (1968), in which the character played by Jack Elam enters a railroad station, pushes the telegraph operator into a closet, and then rips the wires out of a noisy telegraph ticker. Over time audiences expected the wires strung through the fictional western landscape *could* deliver important messages, and yet they also knew that link to civilization was unreliable. If the importance of the telegraph is suggested early in the film, one can expect it to be challenged as the action builds.

The act of wire cutting to increase the anxiety felt by a lonely female under attack was also revised in the horror movies later in the century as the terrified, isolated protagonist makes a call for help, only to hear the line go "dead." Griffith's *Girl and Her Trust* may not let the terror linger or the attack turn to gore, but it is one of the first films to tap into this collective dread of watching a character be suddenly cut off from civilization without any traditional means to call for help.

In addition to showing the tenuousness of wire connections, Griffith's films also mark a shift in the meaning of the telegraph wire strung across a gnarled wooden pole. In 1909 AT&T's temporary takeover of Western Union initiated the slow demise of the telegraph and the rise of the telephone and wireless radio. Ironically, the telegraph emerges as a staple of the American Western just as it is being replaced in the physical landscape. If the absence of a frontier line gave credence to the frontier myth in American culture, then the phasing out of the telegraph in everyday life and the increasing presence of other kinds of wires in the environment enhanced the symbolic importance of the single telegraph line in a vacant landscape. Once upon a time, the wire seems to imply, public safety hung on a single, vulnerable telegraph thread.

The two black-and-white films also speak directly to our own network dependencies and technology's ambiguous role in our daily lives. The two operators sit before telegraph keys at lonely outposts; modern operators and users download new apps, buy the latest smartphones, and have accounts on multiple social networks. Of course, as more telephone, transmission, and distribution lines have crossed the American landscape, our sense of terrifying isolation and vulnerability to technological failure has not been erased. Instead, seeing the lines in these fictional western landscapes can remind us of the possibility that in our real lives, at the worst moment, technology could fail and our secure and civilized towns and cities could suddenly regress into a lawless, ferocious wilderness.

Decades after releasing his first films, Griffith boasted: "Remember how small the world was before I came along? I brought it all to life: I moved the whole world onto a 20-foot screen."[19] The electric telegraph and distribution lines Griffith captured in his films link the small, isolated, lonely world of the past with the expanding city of Los Angeles. In addition, Griffith brought overhead electric lines "to life" by using them as a plot device and encouraging audiences to wonder when that connection between isolated outpost and civilization might be severed. For Griffith and others, however, to "move the whole world onto a 20-foot screen" required more than writers, actors, directors, and sets—it required electric power for lighting, cameras, theaters, and projectors. Some of the electric power that Griffith used to capture and project his narratives onto the silver screen was generated in far-off, truly isolated landscapes.

Between the 1890s and 1910s the telegraph line emerged as a symbol of the tenuously settled American frontier. In western films the telegraph continued to represent technology's *former* advance across American landscapes. Meanwhile, during the same period California utilities built a series of transmission systems that broke world records for voltage and point-to-point distance. Telegraph and power lines shared spaces on the same utility poles, but similar to New York, the popular meanings took different trajectories. The telegraph line reflected the past; power lines projected the future.

Remote Canyons and Sheening Threads

Griffith and the rest of Hollywood could make systematic films about lonely outposts of the West without leaving Los Angeles, but journalists and technical writers traveled outside the cities to dramatize the work of surveyors, engineers, and construction workers building hydropower plants in increasingly remote and harsh environments. In 1912 *Electrical World* called California the "great laboratory of the world in . . . modern energy transmission," an example that the "pioneer spirit is still alive out West."[20] Wooden poles and metallic threads sweeping across the countryside provided a visual intermediary between the "great laboratory" and its "pioneers." The lines had a recognizable form. As electrification spread, what one writer poeticized as the "slender wires bearing their unseen burden" had an evocative presence in the landscape.[21]

California was poised for electric power. In 1879 the nation's first commercial arc lighting system was installed in San Francisco, preceding Thomas Edison's Pearl Street station in Manhattan by three years.[22] In downtown San Jose and Los Angeles wooden towers outfitted with thousands of candlepower arc lamps shone like beacons. Before the "Battle of the Currents" was settled in the East, street lighting systems in Santa Barbara, Visalia, and Pasadena had made their choice by switching from DC to AC steam generators. By 1892 San Francisco, San Jose, Sacramento, and Los Angeles each had electric trains or trolley lines powered by central stations. During the 1880s the growth of electrical grids across the rest of the United States was restricted by the inability to transmit either DC or AC power more than a few miles; in California limited transmission distances and the lack of quality coal and oil deposits to fuel steam-powered generators hindered development.

Hydropower promised to lighten the burden of California's fuel constraints. In 1890 John Wesley Powell—Civil War veteran, explorer, and director of the U.S. Geological Survey—predicted that a system of irrigation canals, dams, and hydroelectric plants stretched across the arid western states could produce "a system of powers . . . unparalleled in the history of the world. Here, then, factories can be established, and . . . the violence of mountain torrents can be transformed into electricity to illume the villages, towns, and cities of all that land."[23] At the time, turbines and dynamos had been used to harness the energy of falling water and to supply power for nearby lighting and mining equipment. The unparalleled potential of California's electric power seemed locked in the restricted courses of distant streams and rivers. Almerian Decker and the advent of long-distance AC transmission opened the figurative floodgates.

A few smaller, private hydropower plants preceded Decker, but the commercial success of his three-phase system initiated a bonanza. In July 1895, forty-foot cedar poles supported cables with 11,000-volt potential from the Folsom plant across twenty-two miles to downtown Sacramento. The electricity powered streetlamps, streetcars, and motors in machine shops, mills, and breweries.[24] This long-distance feat was followed closely in San Francisco, the most populous and, at that time, the most energy-hungry city on the Pacific coast.

The *San Francisco Chronicle* described the "first practical demonstration of the utility and economy of electric power transmission" beneath the headline "Power Sent by Wire: California Pioneered This Field."[25] The *Weekly Chronicle* crowed that California had an "almost unlimited supply of water power" that could be turned into electricity and that the state was now "the pioneer in the field of transmission of power . . . to supply mills and factories with electricity for turning their machinery as well as for propelling cars and lighting streets and buildings."[26] Repeatedly, readers were reminded that the massive generation and consumption of electric power in California was possible because of record-breaking transmissions of power. In this state high-voltage long-distance transmission had evolved from a stage of theory and experiment to demonstrated fact. In 1895 the *San Francisco Examiner* already predicted, "The day is undoubtedly coming when the Sierra water powers can be brought to San Francisco, and when that day comes the city will have an unlimited supply of energy at its command."[27]

This initial frenzy for hydroelectric power was met by some skepticism, similar to that which checked the arrival of tens of thousands of forty-niners expecting to strike it rich in Sierra Nevada streams. One commenter noted that in the spring of 1895, a new kind of "hustler" had appeared on the Pacific coast who brought with him "schemes for the long-distance transmission of electric power" and "talk about setting up water wheels in lonely mountain places and making them give light and cheaply turn other wheels in towns miles away."[28] Hydroelectric power projects did encounter setbacks, including droughts, floods, and lack of funding. Nevertheless, for all the talk about long-distance transmission in the mid-1890s, California entrepreneurs and engineers achieved a series of momentous accomplishments.

In 1896 an 11-kilovolt line connected San Joaquin No. 1 across 38 miles to Fresno, signifying the longest commercial transmission line in the world.[29] In 1899 another distance record was shattered by the 33-kilovolt, 82-mile line from Santa Ana No. 1 to Los Angeles. Superlative lines sending power west from the San Bernardino Mountains and south from the Sierra Nevada would continue to help Los Angeles and its industries sprawl up and down the coast and inland toward the valley.[30] Before the turn of the century the Colgate powerhouse on the Yuba River in Northern California had a 40-kilovolt line spanning 67 miles to Sacramento, and in 1901 the Colgate line was extended 142 miles to Oakland. The hundred-mile mark was superseded again in Northern and Southern California in 1904 with a 60-kilovolt, 242-mile line from the De Sabla powerhouse to Marin County, and a 45-kilovolt, 127-mile line connecting the Borel powerhouse to Los Angeles.

The beauty of Niagara Falls and the sheer volume of water harnessed inside the Niagara Power plant more fully captivated the American imagination, but when the United States joined the Allied forces in the First World War, approximately one-third of the largest hydroelectric, long-distance systems in the entire world were located in California. Eight different transmission lines transmitted at 70 kilovolts or higher across distances of one hundred miles or more: four terminated in San Francisco and four in Los Angeles (figs. 14–16).[31]

The long-distance lines that funneled into San Francisco and Los Angeles began at the figurative mother lode of hydroelectric potential, the California Sierra Nevada. The Sierra Nevada range, which shares its name with a smaller formation near Granada, Spain, stretches approximately 450 miles north to south. In Northern California it runs parallel to the Coastal Range and ends just before the Cascade Range that runs through Oregon. In the south the Sierra Nevada forms the Tehachapi Pass and then flattens into the Mojave Desert, which is abutted by the San Bernardino and San

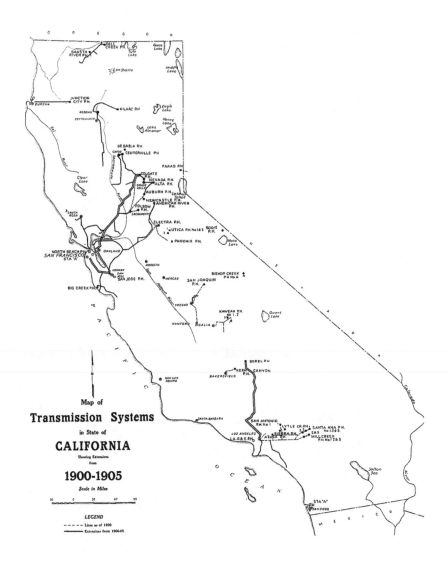

14. Map of transmission lines in California, 1900–1905. W. G. Vincent Jr., "The Interconnected Transmission Systems of California," *Journal of Electricity* 54, June 15, 1925, *Electrical World*, 568.

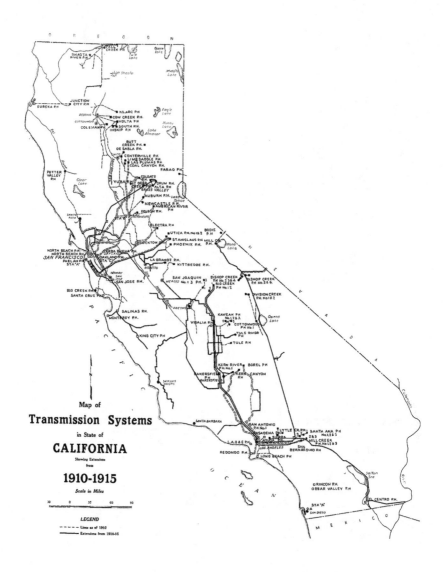

15. Map of transmission lines in California, 1910–15. W. G. Vincent Jr., "The Interconnected Transmission Systems of California," *Journal of Electricity* 54, June 15, 1925, *Electrical World*, 572.

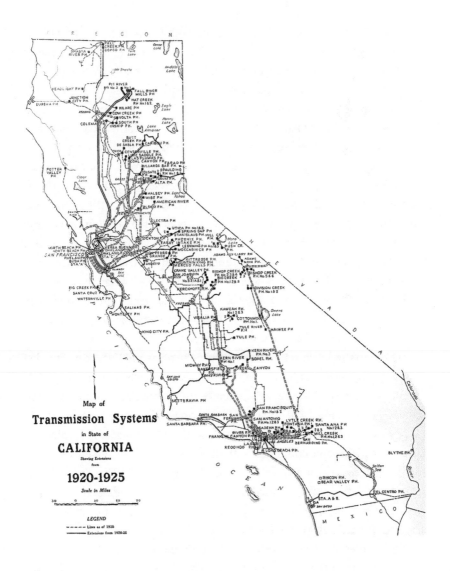

16. Map of transmission lines in California, 1920–25. W. G. Vincent Jr., "The Inter-connected Transmission Systems of California," *Journal of Electricity* 54, June 15, 1925, *Electrical World*, 574.

Gabriel Mountains north of Los Angeles County. At its head and tail the Sierra Nevada curves slightly toward smaller mountains on the Pacific coast to form the Central Valley. Moisture-laden winds from the Pacific saturate the Sierras, and annual precipitation and snowpack, especially in the northern section, feeds thousands of creeks and streams that spread into long rivers—the Feather, Yuba, Pit, Tuolumne, American, Kern, and Owen—and branching river systems such as the Sacramento and the San Joaquin.

During the gold rush, which began near the town of Coloma in 1848, prospectors and mining outfits combed the creeks, canyons, and slopes of the Sierra Nevada and wedged them open with axes, drills, dynamite, tunnels, and high-pressure water cannons. To yield the precious metals and stones fixed beneath riverbeds and crystallized into veins, miners developed hydraulic systems and tools, including dams, ditches, chutes, sluices, and flumes. These high-pressure water systems helped clear away the unwanted soil and rock so the gold and silver could be collected and hauled to market. Innovations in turbine technology, such as the Pelton water wheel, made falling water a marketable commodity across the Sierra Nevada.

Journalists and technical writers viewed the poles and lines that extended from hydropower plants as signs of progress. As early as 1894, an article titled "An Electrical Prophecy" imagines the coming changes to the rural landscape: "Pole lines will be run farm to farm with motors scattered around as needed," and "poles will carry telephone lines for quick communication and the kinetoscope, combined with the phonograph, will enable our gentleman farmer to hear any opera or play he may desire."[32] These poles in the countryside, of course, seemed distinct from those inside San Francisco, where as early as 1903, residents protested plans for a new power line near Dolores Park. Their persistence forced the utility to agree to site the line on taller poles along Guerrero Street and Fillmore.

In contrast to the uneasy response to more overhead lines in the city, the San Francisco–based *Journal of Electricity, Power, and Gas* (later shortened to the *Journal of Electricity*) published ebullient on-site accounts of hydropower plants and their amplifying lines. In many of these articles technical descriptions are supplemented by purple prose. Reporting from the Colgate plant on the Yuba River in 1901, the *Journal of Electricity* waxed lyrical about the plant's extensive operations: "All these poles, and all these lines running out from Colgate present a scene that is probably without a parallel in the world . . . with the aluminum sheening like silver thread in the sunlight, one could almost believe that some giant spider had begun to weave its web, for within the 'parlor' down there somewhere on the river side the ribs of the web radiate out across canyons and over mountains until lost in the sky lines."[33] The lines splintered into every direction, but in some directions, the author adds, the poles "form a forest which . . . seems to be a very dense one." Unlike the ire attached to the "telegraph forest" in New York City or the new lines near Dolores Park, the dense pole forest near the Colgate powerhouse had a commanding presence. These "ribs of the web" that radiated outward were the source of the spectacular changes taking place hundreds of miles away.

In 1901 the sheening silver threads that began at the Colgate powerhouse on the Yuba River made their greatest leap when the Bay Counties Power Company extended the line across the Carquinez Strait to reach Oakland. Enormous concrete anchorages and steel towers on hilltops overlooking the bay secured the cables, which spanned a world record 4,427 feet.[34] The Colgate-to-Oakland line attracted national attention. Eastern newspapers and trade journals reported on the achievement in terms of the technological sublime. The transmission from Colgate and other new technological and engineering advancements made California seem like "a far away country . . . almost out of the world."[35] The *Journal of Electricity* also named the Colgate line the greatest electrical engineering triumph ever accomplished and went on to reference the region's inexhaustible energy potential: "Does the public realize what is meant when one speaks of the 'limitless power' of the Sierra rivers"?[36]

This rhetorical question addresses the difficulty faced by journalists attempting to educate the public about the new, "limitless" sources of electric power. It also hints at why, despite the presence of electric lines in physical and cultural landscapes, this infrastructure has often been considered invisible.

Turn-of-the-twentieth-century ideas about the miraculous and limitless power stored in the West, and especially in California, echoed long-standing

myths of American abundance. From the colonial era Americans had per-
petuated beliefs in the continent's limitless supplies of timber, wild game,
arable land, fossil fuels, and other raw materials. During the Second Industrial Revolution, electric power appeared to be the next bounty bestowed
upon the American people. As Daniel French argues in his book, *When
They Hid the Fire: A History of Electricity and Invisible Energy in America*
(2017), "Limitless on-demand power fit well with long-held American
attitudes about consumption and consequences."[37] Put another way, the
electrical age allowed Americans to amplify their desire for technological
mass consumption with little or no direct personal consequences.

Replacing various forms of fire (e.g., wood piles and stoves, coal bins and
furnaces) with switches, motors, and bulbs that dispense clean, quiet, safe,
instant, and invisible electricity obfuscates our energy choices. Wires make
it possible for our devices to be physically and psychologically detached
from the socket, the meter, the distribution lines, the transmission lines,
and the sources of electric power generation. This is especially true in places
where the fire is "hidden" at coal-fired plants on the outskirts of town or
at distant mine mouths. However, California's initial, striking process of
electrification created distinctive frames for the lines in the landscape.

First, long-distance lines delivering hydropower reoriented public perceptions of the California landscape. Before long-distance power transmission,
California's coastal inhabitants viewed their mountains "as obstacles to the
wagon trains and railroads, as the location of gold and silver mines, and as
the site of winter snows, which, during the spring thaw, brought water for
irrigation but floods as well."[38] The first wave of high-voltage, hydroelectric
transmission lines gave the falling waters of the remote Sierras a visible
presence in farms and factories.

Second, hydroelectric power was acknowledged as being more efficient
and forward thinking than steam-powered systems in the East. One California newspaper explained that hydroelectric power "destroys nothing,"
whereas to generate steam power "some substance is destroyed to supply
the fuel. This means waste."[39] As early as 1902, T. C. Martin—editor of *The
Invention, Researches, and Writings of Nikola Tesla* (1893)—gauged the
economic sustainability of regions provided with geological endowments,

such as coal-rich Pennsylvania, and areas displaying geographic diversity, such as California. Martin concluded, "When the digging of the black coal dwindles, hand-to-mouth diurnal dependence upon sunshine and showers for the 'white coal' that must largely take its place will be the mark of the highest economy and best efficiency on the part of our engineers."[40] A geological survey in 1908 found that 70 percent of the total potential waterpower of the United States was west of Colorado, and 43 percent of the national total was in California, Oregon, and Washington.[41] At the dawn of the twentieth century it seemed likely that the future centers of global industry would have access to cheap, renewable power. This was not a resource to be hidden; it was a possibility to be celebrated.

The notion of a clean, efficient "white coal" also corresponds with the smooth, effortless transmission of power across wires. In 1903 Martin reported how "in the Sierras, the 'downward smoke' of the falling, pausing streams is converted into electric current flashed to the Golden Gate, two hundred miles away, over shining circuits of copper and aluminum filaments."[42] While Niagara Falls had been put in a "harness," the Sierras seemed laced with "shining circuits."

The opening lines of a rhyming poem published in the *Sacramento Union* asked readers, "Do you know that the snow that you cherish is known as the 'New White Coal'?" The poem continues to question if the reader knows:

That its rush is the power of the turbine, which generates the spark,
Which is carried away and away on the wire, dispelling the night
and the dark,
That it drives the busy factory where once the chimney loomed,
And quiets the noisy engine that had puffed and fretted and fumed.[43]

Transforming falling waters into electricity and transmitting the power across great distances helped California cities become cleaner, quieter, and more civilized. Compared to the highly competitive, coal-sourced grids of the Atlantic Seaboard, California's white coal and expansive power lines provided points of pride.

Another distinguishing feature of California's long-distance transmission grid was the simultaneous development of new communities and new industries. In the first decades of the twentieth century California provided a model for rural electrification. By promoting electrical irrigation and water pumping, utilities increased their load factor as the high-energy seasons for farms (spring and early summer) offset peak use in the cities (late summer and winter). In 1924, 23.5 percent of California farms had electricity, compared to approximately 2.5 percent nationwide. In 1940, after extensive efforts by the Rural Electrification Administration, 25 percent of rural homes and farms across the United States had electrical service, but by that time 81.3 percent of California's farms had electrified.[44]

Electrification in California also sparked suburban development, especially near Los Angeles. Henry Huntington, who founded Pacific Light and Power, Pacific Electric Railway, and the Los Angeles Land Company, linked his electric power investments to his transportation systems and real estate business. In 1904 Huntington helped finance the 10-megawatt Borel station on the Kern River. The inexpensive electricity supplied his network of commuter trains, called the "Red Cars," which in turn gave rise to towns such as Huntington Beach and Dolgeville (which was later annexed by Alhambra).

Finally, electrification in the relatively unsettled urban areas of California did not often require ripping up streets to replace gas lighting or retooling factories to receive electric power service.[45] When the Big Creek line was energized in 1913, the *Los Angeles Times* reported, "Electrical energy from the far off Sierras stretched a hand robed with lightning across the gulf of valleys and mountains to the doors of this city yesterday."[46] The power transmitted across these lines supported street lighting, railways, and the fleet of new electrical devices promoted by utilities. They also provided power to emerging industries: automobiles, Hollywood, aerospace, and food packaging. Each of these industrial concerns flourished because of abundant, low-cost electric power. As historian Jay Brigham explains, "People needed water to live, but without electricity Southern Californians would be jobless."[47] In California electrification was not a replacement or supplement for existing industries, municipalities, and infrastructures; it provided the foundation for entirely new ones.

The technological breakthroughs of 1893—AC electrification of the Columbian Exposition, the decision to transmit Niagara power to Buffalo via alternating current, and the successful three-phase AC transmission between Mill Creek and Redlands—each electrified the "new product that is American" envisioned by Turner's frontier thesis. In California that new product was increasingly connected to a vast network of long-distance transmission lines. Newspapers and trade journals celebrated the new lines and the overwhelming positive developments engendered by electrification, yet one reason that the infrastructure may not have been more visible or more fully captivated the public imagination was that between 1892 and 1905 the cedar, sawed redwood, or southern pine pole "product" delivering electric power was often the same, or similar, to those carrying telegraph and telephone lines. The lattice steel tower would be the next electrical innovation to change the material presence of electricity upon the California landscape.

Steel Towers and Spider's Strings

In 1906 the world's first transmission line composed of all steel towers crossed through Newhall Pass, approximately thirty miles north of Los Angeles. In total 1,140 lattice steel structures ranging from thirty and sixty feet tall delivered electricity at an unprecedented 75 kilovolts from the Kern River No. 1 plant to Los Angeles No. 3. The towers rose from the Kern River Canyon, marched southwest across the Bakersfield Plains, and ascended the Tejon Pass. The trail cut through the Tejon Pass for the transmission towers was eventually paved over to make the Ridge Route, then the Grapevine Highway, and in 1970 the eight-lane Interstate 5. Today Interstate 5 is one of the major North-South conduits for the continent; it parallels the Pacific coast from the Mexican border at San Diego to the Canadian border in Washington State. Just as parts of the transcontinental telegraph route of 1861 carved an East-West rut for the transcontinental railroad, which was then followed by Interstate 80, electricity corridors helped to carve out and mark the infrastructural groove between the Sierra Nevada and Los Angeles.

After crossing the Tejon Pass, the Kern River line's towers wound through parts of Piru Canyon, where, as one contemporary article explained, workers had to string lines through narrow gorges and build tower footings on shale cliffs and sandstone ledges. The line then crossed fifteen miles of jagged landscape and cleared the San Fernando Mountains at Newhall Pass, just to the west of the railroad tunnel owned by Southern Pacific.[48] That tunnel, which was completed in 1876, provided a crucial point of entry for the Los Angeles Basin. The Southern Pacific line connected Southern and Northern California. Train cars carrying oranges and other perishable goods, iced and refrigerated by electric power, could now reach Denver, Chicago, and New York in a matter of days when before the journey could have taken weeks. An updated tunnel is still in use, but it has been hidden beneath the bending concrete strips of the Interstate 5–Route 14 interchange. The inaugural steel towers of the Kern River line have long been removed, but taller, higher-voltage replacements carrying electric power from Kern County to Los Angeles still stand sentinel on the western side of Newhall Pass. These steel towers are holdovers from the first high-voltage connection between the Sierra Nevada and Los Angeles and provide another visible layer of the historic conduit for people, goods, and power.

In the early twentieth century lattice steel transmission towers added to the network of railroads, pipelines, and wires that created and sustained what historian Christopher F. Jones calls "landscapes of intensification."[49] These towers physically and visually converged at sites such as the Newhall Pass. These taller, stronger structures could support the heavy strain of thicker cables operating at higher voltages. In addition to helping to make the grid more robust, the towers amplified energy infrastructure's exposure to environmental forces.

Electrical engineers and system designers attempt to limit the lines' contact with nature. They design predictable, efficient, closed systems. The systems are like puzzles, with various malleable and dangerous pieces—generators, switches, bus bars, transformers, and transmission lines. Exchanges between these pieces and the external environment are intentionally limited, even undesired. As the amount of power going into the system and the diversity of uses increases, solving this puzzle tends to move physical infrastructure higher into the sky and farther from public view.

For example, when California Edison finished the Kern River–to–Los

Angeles line, *Electrical World* noted that the slate-colored pin-type insulators "harmonized" with the metal hue of the towers. The gray insulators, Victor M-4800s, did not, "afford as prominent a target for the malicious marksmen as do those of ordinary brown glaze."[50] In 2003 two insulator collectors scoured Pico Canyon and recovered one of the sixty-seven-pound artifacts, the "largest [pin] insulator used for long distance transmission."[51] These insulators are a rare find today, but for contemporaneous engineers the insulators were notable for their ability *not* to be noticed and subsequently used for target practice.

Environmental historian Etienne Benson has described another important instance of the "insulating" actions taken by Southern California Edison to protect the lattice steel transmission towers. In 1923 SCE raised the voltage on a 243-mile line from Big Creek to Eagle Rock from 150 to 220 kilovolts. These lines used suspension insulators, sometimes called "accordion insulators," for which a string of smaller insulators hangs from the tower. In the early twentieth century suspension insulators replaced the pin-type insulators for lines operating above 33 kilovolts. For the Kern line upgrade the increased voltage also required hoisting the towers higher into the air to gain greater ground clearance. Workers used jacks to lift the existing towers and to insert new bases. The work was completed with the lines still energized. In the months following the upgrade, a series of unexplained flashovers threatened the reliability of the new 220-kilovolt line and the broader power system. Harold Michener, an electrical engineer and amateur ornithologist, traced the flashover problem back to the conductive, streaming effluents produced by large birds as they took flight from the towers, where they perched to look for prey.[52]

The steel towers were more weather resistant, but they were also more attractive for red-tailed hawks seeking resting places and launching spots from which to hunt. Therefore, Michener and his colleagues designed spikes that could deter birds from perching on the crossarms and pans to collect the strings of excrement before they touched the energized lines. They also added automatic relays, so if and when flashovers did occur, the load would be instantly switched to another line, minimizing any disruption in service.

This was neither the first nor the last attempt to prevent disruptions caused by animals to the electric transmission systems. In the 1920s cedar poles and fir crossarms were treated with creosote to deter termites. More recently, a series of blackouts have been traced back to squirrels.[53] Benson suggests such acts of insulation and interconnection used to mitigate animal impacts or unexpected weather tend to "make nature invisible" for consumers.[54] Southern California Edison engineers desired transmission infrastructure to be impervious to the point of being invisible to marksmen, hawks, and consumers, but their marketing teams attempted to capitalize on the increasingly familiar sight of steel towers.

Southern California Edison used outlines of electric power infrastructure to promote the technical, political, and financial unification of its operations. Managers and executives recognized that their company image was shaped by the quality of their service as well as by the image conveyed by powerhouses, substations, and power lines. In materials such as pamphlets and maps, the transmission tower acts as a synecdoche for the power company.

Historical documents show the lattice steel tower in positions of strength and conveying uniformity and reliability. In 1923, the same year SCE engineers made their modifications to the Big Creek line, a company newsletter explained, "What we have to sell is invisible, but on every hand may be seen the apparatus whereby our product is marketed." The newsletter continues: "We now have 540 miles of 220,000 volt tower lines. To serve 6 million [the anticipated population of their service area], will require 3,000 miles of such lines."[55]

In other instances *Southern California Edison* is printed in relation to a single transmission tower without wires, or the name hangs over a line passing through the landscape. Utilities across the country used sketches or drawings of lattice steel transmission towers in company emblems and letterhead; SCE also inserted such images in its marketing campaigns. A pamphlet released in 1931 invites investors to "Follow the Wires . . . to Financial Independence" (fig. 17). Rather than overlooking or ignoring the lines in the landscape, customers were encouraged to "follow" them as part of a long-range investment plan. On the inside page another end point for this investment journey is revealed: "Follow the wires . . . and see a few of the Three Hundred & Fifty Million Dollars' worth of assets of the Southern

California Edison Co." The wires provide power to the general public; for stockholders the lines symbolize an investment operating in the landscape.

In addition to attracting investors, SCE used high-voltage transmission lines in narratives that merged pioneer rhetoric with visions of modernity and progress. In 1917 SCE settled a long-standing suit with the Los Angeles Department of Water and Power and also gained control of Pacific Light and Power, which had built and previously operated the Big Creek system. SCE was now the fifth largest utility in the United States. That year the company released a fascinating advertisement that included an Adonis-like figure holding a transmission tower. The caption reads: "In the vanguard of the pioneers who have made California one of the garden spots of the world, we find the electric utilities. True pioneers—the electrical utilities have harnessed our water powers to meet consistently the growing needs of a greater California" (fig. 18).[56]

The drawing is a crude copy of John Gast's *American Progress* (see fig. 6). A wagon train in both images crosses from right to left as a gigantic figure hovers overhead. In Gast's version a telegraph line slips from the fingers of the angelic "Star of Empire" as she floats west, civilization at her heels. In the modernized rendition the male Adonis holds a lattice steel tower in his left hand and seemingly converts its energy into the beam of light extending from his right. The landscape does not include other technological icons such as the railroad. The Adonis figure, with its shining beam of light, guides the settlers through this barren land. As the caption implies, this is not an homage to settlers and homesteaders who used technologies such as the ax, rifle, railroad, and telegraph as they made their way across the continent; this is a corporate advertisement for the so called true pioneers, the companies and engineers who harnessed electricity so that subsequent generations would be able to enjoy the comforts of electrification in California.

Two concurrent developments of 1931 encapsulate the changing inflections of electricity and landscape in the age of the lattice steel transmission tower. First, an SCE pamphlet titled *Edison Facts* includes a photograph taken by Haven G. Bishop from Carrack Avenue near the industrial waterfront in Long Beach (fig. 19). The black-and-white photo shows three towers

17. "Follow the Wires,"
Southern California
Edison advertisement.
Box 385 (7), Southern
California Edison
Archive, Huntington
Library, San Marino.

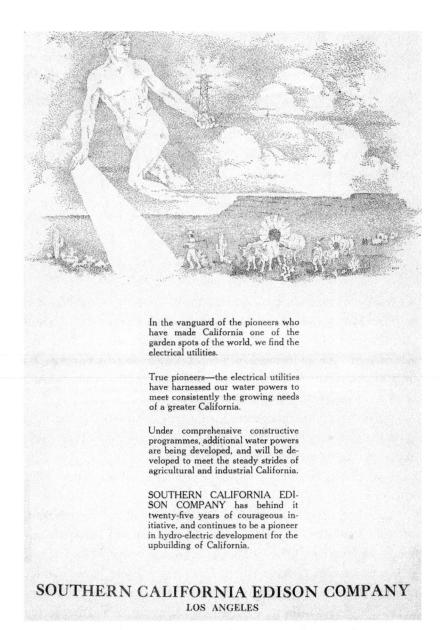

In the vanguard of the pioneers who have made California one of the garden spots of the world, we find the electrical utilities.

True pioneers—the electrical utilities have harnessed our water powers to meet consistently the growing needs of a greater California.

Under comprehensive constructive programmes, additional water powers are being developed, and will be developed to meet the steady strides of agricultural and industrial California.

SOUTHERN CALIFORNIA EDISON COMPANY has behind it twenty-five years of courageous initiative, and continues to be a pioneer in hydro-electric development for the upbuilding of California.

SOUTHERN CALIFORNIA EDISON COMPANY
LOS ANGELES

18. "The Adonis Pioneer," Southern California Edison advertisement, 1917. Box 383 (11), Southern California Edison Archive, Huntington Library, San Marino.

that form part of the Long Beach–Lightpipe–Laguna Bell line. Rising to a height of 310 feet, these towers were at one point the tallest transmission towers in the United States. Today they support forty-two cables crossing the Cerritos Channel. In the photograph the stacks of the Long Beach Steam Station on Terminal Island appear in the background.

Beneath the photograph are the lines "Draw with idle spiders' strings / Most pond'rous and substantial things." The complete phrase, which comes from Shakespeare's *Measure for Measure*, reads:

How many likeness, made in crimes
Making practice on the times
Draw with idle spiders' strings
Most pond'rous and substantial things.

For those who owned and controlled the material lines that branched out from hydropower stations set in the high Sierras and connected to the fuel-powered plants such as Long Beach and Redondo, these "spiders' strings" seemed like neat, geometrical, and substantial things. The simplicity of the strings certainly concealed the political and economic advantages wielded by the utilities and the various regulatory agencies and governing bodies that controlled the flow of power. Starting in the 1910s, the American public became increasing leery of the "spider" webs of power and the entities that owned, controlled, and sited them. The fierce fight to regain public control over electrical infrastructure in Los Angeles was certainly substantial.

Meanwhile, also in 1931, the *New Yorker* published a profile on Henry Dreyfuss, an upstart industrial designer. Someone at Bell Telephone Laboratories had recently presented the twenty-seven-year-old Dreyfuss with an insurmountable design problem: "How to make telephone poles less unsightly?" The *New Yorker* piece included the parenthetical quip "(he hasn't solved that one yet)." Thirty years later Dreyfuss went beyond the problem of unsightly telephone poles and designed new high-voltage towers. These towers had the potential to permanently reconfigure California's visual engagement with power.[57]

Draw with idle spiders' strings
Most pond'rous and substantial things
—*Shakespeare*

19. Transmission lines crossing Cerritos Channel, photo by Haven Bishop, 1931. Bishop's original photograph was republished and captioned with the quotation from Shakespeare on a pamphlet titled *Edison Facts*, August 1936. Box 381 (8). Southern California Edison Archive, Huntington Library, San Marino.

The Limits of Aesthetic Transmission Line Design

When a route becomes routine, the landscape often scrolls into unconscious-ness, and for many drivers, roadside infrastructure is banal enough to be entirely dismissed. Even artists hoping to draw poles and wires can have difficulty paying attention to them. California artist Robert Crumb kept a reference album of utility pole photographs so he could re-create them in the backgrounds of his drawings. Photographic evidence was necessary because, as Crumb bluntly explained: "You can't make up this crap; you know it's too complicated. In the real world this stuff is not created to be visually pleasing. It is the accumulation of the modern industrial world. People don't even notice it. They block it out." In Los Angeles millions of daily commuters block out the billions of complicated wires, poles, and metallic gadgets they pass over, beneath, and alongside. Yet those who choose to pay attention to the electrical lines stretched across the city may notice a few structures that stand out from the rest of the wiry "stuff." These are some of the first, and last, aesthetic transmission structures erected in the United States—the Dreyfuss designs.

Between 1964 and 1968 Southern California Edison manufactured at least three types of designer power lines. One was the "portal" pylon, an example of which flanks either side of Highway 101 through Conejo Pass. Another was the "Starburst" design for 69 kilovolts and which can be seen along Holly-wood Boulevard. The Starburst utilizes a familiar-looking wooden pole, but instead of horizontal crossarms, it has six cantilevered insulators that radiate from a point near the apex. The final is the "Sunburst." The Sunburst struc-tures reflect one of the strongest applications of the midcentury modern aes-thetic to electric infrastructure. Utility executives, engineers, and industrial designers believed that these clean, streamlined, integrated structures could lessen the public's distaste for transmission lines. In 1967 two of the Sunburst specimens were set within a row of lattice steel towers at Huntington Beach in order to test their strength and public reaction. On both accounts the towers earned favorable marks, and in 1968 thirteen more Sunbursts were erected along a 1.2-mile easement owned by Standard Oil Company to com-plete an expansion of Rosecrans Avenue in El Segundo (fig. 20).

The Sunburst design includes two light-gray tubular steel poles. The bottom is similar to an A-frame setup, but in the upper section each pole is angled so as to appear directly vertical and parallel. Three black pointed crossarms span the gap between the two 140-foot tall steel poles. The middle crossarm is horizontal; the top and bottom arms are slightly angled. An SCE executive calculated that a Sunburst tower cost approximately 30 percent more than a lattice steel structure, but he also noted the extra cost could be recovered if the utility avoided delays, rerouting, legal battles, and messy public relations campaigns.[58] Local newspapers and civic officials offered positive comments on the appearance of the towers, and in 1969 Dreyfuss and Robert Coe accepted the nationwide "Design in Steel" award for these "futuristic," "eye-pleasing," and "ultra-modern" structures.[59]

A handful of the Dreyfuss designs were adopted elsewhere in the world—an interesting modification of "jolly green giants" delivered power to Disney World in the early 1970s.[60] Still, the Dreyfuss designed structures on display in El Segundo and other parts of Los Angeles are anomalies.[61] The most common adaptation of the original designs is a brown version of the Sunburst. One of these unfortunate variations carries wires across Interstate 210, about thirty minutes south of Newhall Pass. The two parts of the A-frame seem to fall toward one another like poles in a teepee. In addition, the original sleek gray and black parts have been replaced with a khaki color that neither accentuates the tower's form nor effectively blends into the landscape. Finally, these brownish Sunbursts often share rights-of-way with smaller distribution lines or lattice steel towers. The juxtaposition of taller Sunbursts and shorter wood poles erases any positive aesthetic impact that the original Sunburst design may have conveyed in a more open and spruce setting.

Early in the century power companies sited and designed transmission lines based upon the limits of engineering and the bottom line. Poles were chosen for cost, height, strength, and reliability. The first line routes were, according to one executive in California, determined by "rather simple economics," which meant the poles were erected on lands where they could function properly—highways, railroad rights-of-way, fields, and ranches— and which required the lowest financial cost. Initially, Californians accepted

20. The "Sunburst" towers designed by Henry Dreyfuss and Associates. Photo by Art Adams, 1968. Southern California Edison Archive, Huntington Library, San Marino.

the lines that ranged across these relatively desolate landscapes. In 1902 *Electrical World* noted that in California "everybody working for the common good" meant that "nobody raises an eyebrow at the installation of a 25[-] or 35-mile transmission of 15,000 or 20,000 volts." By contrast, in the East "even a very moderate transmission stirs up opposition with a sharp stick."[62] As the century progressed, pole lines serving the California "common good" intersected more desirable areas. The more visible lines were decried as having various and increasingly severe environmental impacts.

Architect E. H. Bennett voiced early concerns with the wooden pole's impact on the California landscape. By 1925 the American city, Bennett observed, had been "rendered hideous by poles of every kind," and the blight would soon extend into the countryside. Along the California coast, for example, "long stretches from the mountains to the shore" made pole lines more obvious. Unlike the trees that helped to relieve the flatness of this desert scenery, the "thin outlines" and "ragged appearance" of California's utility poles meant the lines appeared to "sprawl in all directions." Situating pole lines and substations based on topographical analysis and partially disguising them with trees, hedges, and shrubs could improve their visual impact and create, Bennett suggested, "landscape rendered more interesting because of their presence."[63]

Bennett railed against wooden poles and inconsiderate siting techniques; he also implied that transmission towers, if properly placed, could achieve "a grandeur even on hillsides." There was, Bennett claimed, "at times something irresistibly fine in the aspect of great airy structures stalking the hills."[64] Such comments are representative of the technological sublime sometimes attached to transmission towers in the 1920s and 1930s. As more steel pylons joined and replaced wooden utility poles, filmmakers, painters, and poets drew new inspirations—and fears—from the geometric structures marching through the landscape. The British school of "pylon poets," including W. H. Auden and Stephen Spender, viewed the new lattice steel towers of the National Grid as powerful, rational, and unifying symbols as well as omens of a future marred, as Auden describes, by "Pylons falling or subsiding, trailing dead high-tension wires." Spender's "Pylons," published in 1933, presents these "outposts of the trekking future" striding unheeded through the newly electrified landscape, displacing pastoral sentiment with "these new-world, rational towers."[65] (Interestingly, the Milliken Brothers Company of the United States designed the "new-world, rational towers" that revolutionized the shape and perception of the British landscape.)

In the 1930s poets celebrated and subverted the image of pylons that soared hundreds of feet in the air. The challenge Dreyfuss confronted was how to balance the widespread desire for uncluttered California scenery with the utility industry's need to build taller, higher-voltage transmission lines to satisfy growing energy demands. As biographer Russell Flinchum notes, Dreyfuss's task amounted to making "the necessary invisible, and when visible, attractive."[66] To find this balance between form and function, aesthetics and engineering, Dreyfuss and his team envisioned transmission towers as art forms.

Dreyfuss was comfortable repositioning the values of aesthetics and functionality within the parameters of art, design, engineering, and infrastructure. He was born in Brooklyn in 1904 and grew up working in his father's costume and theater supply shop. During high school he apprenticed for Norman Bel Geddes, and at age nineteen he was hired as artistic director for the Strand, a three thousand–seat auditorium at Broadway and Forty-Seventh Street. For a starting salary of fifty dollars a week, Dreyfuss designed and managed equipment, props, costumes, scenery, and lighting for the Strand's rotating bill of short films, vaudeville acts, and live music. In 1924, when D. W. Griffith released his last major motion picture, *America*, Dreyfuss likely built the set upon which it was projected.[67] At one point in 1930 five different Broadway shows featured Dreyfuss set designs.[68]

In his midtwenties Dreyfuss transferred his experience designing fictional worlds to making objects of mass production. His pivot between professions suggests that if "all the world's a stage," then great architects and designers understand how to repurpose props and scenery. Dreyfuss's systematic studies of human movement and engagement with devices and spaces—what is now called "ergonomics"—led to hundreds of domestic products sold by Macy's Department store and famous household devices including Western Electric Trimline and Touchtone telephones, Honeywell thermostats, and Hoover vacuum cleaners. Dreyfuss also designed massive, user-friendly interior spaces including movie theaters, passenger trains, airplanes, transatlantic ocean liners, World's Fair exhibits, and "the great strategy room for the Joint Chiefs of Staff in which a large part of the [Second World War] was planned."[69]

When the Second World War ended, Dreyfuss and Doris Parks, his life partner and business manager, relocated their primary residence from New York City to South Pasadena. They maintained offices on both coasts and served an international clientele during the years of postwar growth and prosperity. Meanwhile, between 1945 and 1960 total use of electric energy in the United States tripled, from 275 to 848 megawatt hours.[70] The increase in electric power was accompanied by the further proliferation of high-voltage lines and lattice steel towers. In 1963 approximately ten thousand miles of lines with potentials of between 345 and 500 kilovolts crossed the country.

During this period of intense energy development and rising environmental consciousness, the popular perception of overhead distribution and transmission lines swung from technological triumph to industrial blight. Subdivisions, strip malls, and freeways each seemed more acceptable parts of the landscape than overhead transmission lines. Citizen groups such as California Beautiful and Los Angeles Beautiful (of which Dreyfuss was a member) argued that wooden poles and low-hanging wires made unnecessary intrusions on outdoor scenery. Utilities tended to sympathize with such complaints and attempted to consolidate pole lines or, when feasible, to bury distribution lines. As one executive for Southern California Edison explained, "Under grounding has sort of taken the place of mother love—everybody's for it and hardly anybody dare be against it."[71]

Distribution lines, especially in new housing developments, could be buried with relatively reasonable expense. For areas that required high-voltage transmission lines, this conciliation was often untenable because of the astronomical costs of construction and maintenance. To appease opponents, utilities attempted to site new towers and substations away from public view or block them with trees or other shrubbery. A 1965 report describes how the "hornet's nest of public opposition" inspired utilities to "accommodate the consumer's outraged esthetic sensibilities by camouflaging the occupancy of electrical facilities."[72] The camouflage was not effective. Citizen groups and environmentalists continued to demand that distribution and new high-voltage transmission lines be moved to other areas or placed underground.

Aesthetic critiques latched onto the lines in the California landscape and sparked public outrage. One case of vigorous resistance occurred in 1965, when the U.S. Atomic Energy Commission (AEC) planned a 220-kilovolt line to connect the Pacific Gas and Electric grid to Stanford University's new $114 million linear accelerator. In an attempt to negate the utility's plans, the city of Woodside passed an ordinance banning construction of any new overhead transmission lines. The AEC countered by agreeing to use shorter poles instead of lattice steel towers and to paint the poles green to help them be less visible. The community refused, and an initial court ruling upheld Woodside's overhead power line ban. Eventually, the federal government stepped in and approved the line with green poles. The citizens were irate, but the pleas for aesthetics, beauty, and local autonomy had been overcome by federal organizations and what the *New York Times* called "high-powered bureaucratic arrogance."[73]

In the 1960s the need to expand power grids and the public's desire to bury wires convinced SCE to redesign its overhead transmission lines.[74] Vice President Robert Coe spearheaded the effort and enlisted Jordan Lummis as the consulting engineer. Loomis lived approximately two miles from Henry Dreyfuss, and he asked his friend to help with the new designs.

The result of the initial collaboration between Dreyfuss and Lummis was the Skyburst and Sunburst designs. Positive reactions to these new designs in California inspired Edison Electric Institute (EEI) to further examine whether or not aesthetic designs might tame public resistance in other areas. In 1966 the EEI secured $400,000 for Dreyfuss and Associates and a team of engineers to imagine an entirely new fleet of lines from 40 to 365 kilovolts. Specific structures for city, suburbs, and rural areas would result in what the *New York Times* anticipated as "prettier power lines."[75] For many onlookers the program seemed gratuitous. The *Wall Street Journal* announced that "beautification fever" had finally forced utilities to improve the appearance of "those gaunt, steel structures resembling the skeletal remains of giant praying mantises."[76] *Architectural Forum* applauded Dreyfuss's design efforts, but it did not enthusiastically embrace the utility's "pacification program."[77]

Dreyfuss displayed mixed feelings about the project's results. In 1966 he told a local reporter that undergrounding all lines was impractical and, therefore, "what we must do is to make the utility pole into a piece of sculpture, and to make high voltage lines into an art form." The following year Dreyfuss made it clear that he might design transmission structures, but he would not be responsible for building them. In other words, any final product set into the landscape would not be *his* art form. "For the first time in our professional career," he told *Electrical World*, "the end product is not going to be a product at all. It is going to be a large book." Dreyfuss may have distanced himself from any physical manifestation because he had built his career by undertaking careful studies of human interaction with objects and environments. Theater stages, train cars, and home appliances could be crossed, entered, or touched. For Dreyfuss good design offered a smooth, consistent, reliable, pleasurable human experience. Even a bridge could be traversed and therefore seemed more engaging and less impersonal. As Dreyfuss clarified, "We use the electricity on a transmission line, but physically, we don't use the tower itself." The inability to "use" the tower put Dreyfuss in a difficult position, and it seems fair to ask whether or not Dreyfuss believed public attention should be directed toward transmission lines or if he viewed his final designs as artworks.[78]

Dreyfuss's uncertainty regarding the beautification of transmission lines is understandable. In some instances Dreyfuss explicitly stated that the best possible towers would be "as unobtrusive as possible, perhaps they won't even be observed or noticed."[79] Dreyfuss repeated this desire for infrastructural invisibility in the introduction to the final book of the *Esthetic Designs for Transmission Structures* project. He writes: "All of us would prefer to place electric power underground or to transmit the power on an invisible beam. But until such techniques prove practical, we hope that the work shown here will serve usefully those dedicated designers and engineers who seek to preserve the integrity of our surroundings."[80] The best way to preserve the *visual* integrity of the landscape, it seemed, was for the public not to see power lines on the landscape. Second best would be if the public did not have to see something as ugly as the traditional lattice steel tower.

Despite his desire for invisible transmissions and unobtrusive infrastructure, Dreyfuss entertained the various forms and functions for these artifacts. As any standardized design might appear repetitive if placed one after another into a diverse landscape, Dreyfuss thought of "doing two designs—two variations on a design and then alternating them . . . two-one-two, like a piece of music." He also imagined: "We can use the base of some of the towers for shelters, even playground equipment. Slides on the bottom of the towers would be fun." Dreyfuss dreamed of towers that could merge the functions required by the utility—safety, strength, durability, easy access and maintenance—with those preferred by the public, recreation and aesthetics. This goal is made clearest in the introduction to *Esthetic Designs for Transmission Structures*, in which Dreyfuss writes: "Today we proudly show visitors our bridges as scenic wonders—from the stately George Washington in New York to the harp-stringed Golden Gate in San Francisco. When transmission towers are given the same purity of expression given great bridges, they, too, may be acclaimed as a Twentieth Century art form."[81]

The ensuing pages attempt to show the range of functional, aesthetically innovative structures. Photographs of a tower model with small figurines are juxtaposed with detailed drawings, schematics, variations on the design, and suggestions for materials, including plastics, aluminum, and concrete. Toward the end of the book the authors explain the potential "two-one-two" patterns that might be used to create a different visual rhythm on routes with a long, continuous exposure.

Dreyfuss's hesitant approach was necessary, yet in a different social and technological landscape it seems possible that this book could have been a blueprint for sweeping changes to the American landscape. Southern California Edison tried to convince the public of the lines' aesthetic potential. As Lummis later explained, "Appearance is purely a subjective matter so opinions of beauty can be swayed by words."[82] In the 1972 SCE manual *Aesthetic Guidelines for Electric Transmission Lines*, the authors argue for the positive impact Dreyfuss designs could make on the Southern California landscape. A photograph of a Sunburst against a sunset is accompanied by the caption: "The impression to the viewer is neat and orderly. Because of its nearness to the observer, and the sparseness of Southern California greenery, it cannot be hidden. Hence, its impressive size is made more dramatic, in this instance, by the simplicity of the silhouette" (fig. 21).

Ultimately, neither the aesthetic designs nor the words used to describe them eased public complaints about the visual impact of power lines. The collaboration between Edison Electric Institute and Dreyfuss and Associates was "the first and the last cooperative attempt by industry to create new aesthetic structure designs."[83]

Almost a decade later, a *New York Times* editorial titled "There's No Way to Beautify a Power-Line Pylon" referenced the seemingly forgotten Dreyfuss designs to prove that any attempt at improving aesthetics, "however well designed, will not achieve the only desirable goal, that of invisibility."[84] As Dreyfuss implied and critics continued to suggest, the best the public could hope for was to have the lines removed as far from view as possible and, when they had to be visible, attempt to reduce the visual impact.

California Lined by Power

Between 1905 and 1945 hydropower produced 60 to 80 percent of the electricity coursing through California's flourishing, state-of-the-art, interconnected power grid. In these decades California came closer than any region on the planet to achieving Tesla's turn-of-the-century prophecy of harnessing industrial-scale energy from waterfalls and other sources that he said were "forever inexhaustible" and then providing widespread public access to electric power. Of course, the West's hydropower resources were not boundless. Increasing demands for energy, technological advancements (especially those related to nuclear fusion), lower fuel costs, and the pressure to improve system reliability drew more coal, natural gas, and nuclear power into California's energy portfolio after 1945. Furthermore, Tesla wanted to transmit high-voltage currents between a system of wireless towers, and some say he planned for the electricity to be free.[85] Hydroelectric transmission in California was relatively cheap but never free, and it required tens of thousands of miles of high-voltage transmission lines.

21. Henry Dreyfuss and Jordan Lummis, 1967. Dreyfuss (*left*) and Lummis (*right*) were friends and neighbors who worked together on transmission line designs in the 1960s. Here they examine a model of the Sunburst towers. In a tragic biographical coincidence, in 1972 Doris Marks, Dreyfuss's life partner and frequent collaborator was diagnosed with terminal cancer, and the couple chose to end their lives together in the garage of their family home in South Pasadena. Twenty years later the ninety-one-year-old Lummis and his wife, Gregory, made the same decision and were found dead of self-inflicted gunshot wounds in their home a few miles from the Dreyfusses. Lummis, according to a June 23, 1991, article in the *LA Times*, had recently told a neighbor he was getting "too (expletive) old. It's just no fun." Photo by Joseph Fadler, 1967. Southern California Edison Archive, Huntington Library, San Marino.

In the 1920 and 1930s the lattice steel transmission tower replaced the wooden utility pole as an icon of technological progress on the American landscape. Both the tower designs and pole configurations changed slightly between systems and regions, but the silhouette, lattice form of these towers was featured on letterhead, pamphlets, advertisements, and system maps. For Southern California Edison the transmission tower stood for electrification's revolutionary effects and provided a symbol of the grid serving its customers.

In the 1950s abundant, seemingly unlimited, deregulated electric power saturated American life. The banal poles and chaotic wires became more commonly viewed as symbolic of a dangerous blight on otherwise aesthetically appealing landscapes. Southern California Edison attempted to resolve the problem of power lines' aesthetic impacts with new designs such as the Starburst, the Sunburst, and the Portals that stand above Conejo Pass.

As telegraph, telephone, and power engineers wired the California landscape the infrastructure sometimes symbolized the conquering of distance and the ability to reliably satisfy energy demands. Over time public perceptions shifted, and power lines were viewed as dangerous eyesores. The Dreyfuss designs represent an attempt to mediate between the visible and invisible as well as sublimity and blight; no such mediation would have saved Southern Californians from the costly debacle surrounding the Tehachapi Renewable Transmission Project and the "monster" towers that invaded Chino Hills, thirty-five miles east of Los Angeles.

4

Public Perceptions and
Power Line Battles, 1935–2013

At first the power line battle in Chino Hills seemed like another episode in the bitter, sometimes violent struggle to route, or reject, overhead transmission lines. In 2009, despite numerous legal motions and thousands of public complaints, the California Public Utilities Commission approved California Edison's preferred route for a 500-kilovolt overhead transmission line through the middle of a suburban neighborhood. The commission felt this was the "environmentally superior" route. The line would occupy an easement Edison owned and had controlled since the 1950s, when everything around it was ranchland. Houses now flanked the same right-of-way, but Edison could not be blamed for hasty housing development or limited city planning. It had a right to protect its investment.

Two years later Edison construction crews used large cranes and helicopters to erect lattice steel towers and tubular steel poles and to hang glass insulators. Fourteen of the new towers occupied a 3.5-mile stretch of Chino Hills. Approximately one thousand homes would now be within 500 feet of the new towers. Citizens who had previously shown little interest in Edison's plans now had visible evidence of the impact and could imagine the effect of adding six high-tension cables to the existing structures. Local opponents shared photographs of the towers and made YouTube videos such as "The Monsters Have Arrived," "The Haunting," "Mutilation of Chino Hills," and a catchy music video, "Pack It Up, Get It Out, s-c-e!" The 1,500-citizen strong grassroots organization Hope for the Hills rallied beneath the tower at Coral Ridge Park and painted large yard signs making its case against what its members saw as the poisonous, unethical, scandalous transmission line and its nefarious owner, Edison.

Hope for the Hills members published editorials, wrote letters to elected officials, and organized via social media. Some took pride in being blocked by the managers of Edison's Facebook page for making antagonistic comments. They stamped their profile pictures with BANNED in red letters and a strip of digital black tape across their mouths. During a pivotal few months hundreds of these vocal homeowners wearing their trademark bright neon shirts traveled to city council meetings, Edison board meetings, and public hearings of the California Public Utilities Commission. Years later Bob Goodwin, president of Hope for the Hills, told me that a few members "went rogue" and sent coal and rubber rats to the homes of Edison executives via certified mail. Edison used advanced engineering and legal arguments to push a particular route for its renewable transmission line; local homeowners took equally extraordinary measures to stop it.

The struggles to plan and build new transmission lines through suburban and rural landscapes can be gleaned from reviews of pamphlets, advertisements, and documentary films supporting public power in the 1930s and 1940s as well as from news reports, exposés, and made-for-TV movies exploring power line battles from 1970s to the present day. Together these cultural artifacts reflect a broader, symbolic transition of overhead lines from progress to necessary infrastructure to burdensome blight. Social scientists continue to research (and potentially mitigate) the correlations between public perceptions and the likelihood and force of public opposition. This research may benefit by more fully considering ways that the aesthetic experience of landscape shapes power line debates.

Rural Lines as Beacons of Freedom and Isolated, Affluent Complaints

In the first decades of the twentieth century, utilities and their corporate counterparts created massive advertising campaigns to sell electricity as the clean, productive, smart way to live. After the First World War production and consumption continued to rise, and electrification slowly expanded to suburban and rural markets. During the 1930s, at the height of the Great

Depression, private sector power generation increased by nearly 46 percent, and federal power generation increased twentyfold.[1] Total domestic consumption of electricity nearly doubled.[2] Although the majority of electric power was still consumed in urban and industrial centers, the rise of public power and the electrification of millions of family farms had far-reaching consequences for the utility industry and rural America.

For much of the early twentieth century, small towns, agricultural areas, and Native American reservations remained in the dark. In 1933 only 11 percent of farms had electric service. That year Franklin Roosevelt initiated the Public Works Administration to help finance various power and lighting projects. Two subsequent federal actions helped to initiate change. First, the Public Utility Holding Company Act of 1935 gave the Securities and Exchange Commission and Federal Power Commission more authority to regulate the utility industry. Massive utility holding companies had emerged during the 1910s and 1920s. By 1930 nineteen holding companies managed 90 percent of all investor-owned utilities, and in 1932 eight of them controlled 75 percent.[3] Through complex mergers and leveraging techniques, the holding companies inflated book values and disguised losses. In the nineteenth century economies of scale and natural monopolies helped powerful railroad and telegraph companies expand across the country. In the twentieth century utility holding companies took advantage of the same economies of scale as well as the general lack of restrictions for massive, neoliberal conglomerates. The holding companies pioneered new accounting schemes, debt and property swaps between subsidiaries, and insider trading.

Samuel Insull's Middle West Utilities Company was an infamous offender. Insull had been Thomas Edison's secretary when they opened the Pearl Street Station in Manhattan. In 1892 Insull moved to Chicago to develop Edison's interests. Over the next three decades he built the blueprint for the investor-owned utility of the twentieth century: centralized coal-powered stations, tiered and metered rates, exclusive access to the market, and a predictable and diversified load. After taking over Chicago's power grid, Insull applied a similar technological and managerial system across the country. Before the Stock Market Crash of 1929, Middle West owned $27 million in capital to control $500 million in assets across thirty states and fifty-three hundred communities. After the crash what appeared as a $2.9 million surplus on the books was calculated as a $177.7 million deficit.[4] The collapse of Middle West Utilities left many smaller cities and towns to endure years of unreliable service and unpredictable rate changes. Federal prosecutors charged Insull with mail fraud and breaking antitrust laws. He fled to Italy and Greece before being apprehended in 1934 and extradited to Chicago, where he was found not guilty on all charges. Four years later Insull died of a heart attack on a Parisian subway platform. While Orson Welles's *Citizen Kane* (1941) is often seen as a pseudo-biography of the media magnate William Randolph Hearst, Welles wanted Charles Foster Kane's physical demeanor to match Insull's and later said: "There's all that stuff about [Robert] McCormick and the opera. I drew a lot of that from my Chicago days. And Samuel Insull."[5]

The New Deal and the federal oversight of investor-owned utilities concurred with the creation of the Rural Electrification Administration (REA). President Franklin Roosevelt initiated the administration in 1935 and signed the Rural Electrification Act in 1936. Nebraska senator George Norris, one of the program's leading architects, had seen firsthand the "grim drudgery and grind" that plagued generations of homesteaders. For Norris rural electrification equaled emancipation, and public power helped farmers control their fate. Technologies and systems such as the plow, irrigation, and the railroad had been instrumental in the success of homesteaders, especially in the Great Plains. According to Norris, the American farmer, due to his greater appreciation for useful technologies, would make a "more satisfactory consumer of electricity than the individual in the town and city."[6] The REA offered a framework for small towns and groups of farmers to finance, build, and operate transmission and distribution systems. By 1950 the REA had helped electricity reach 79 percent of American farms.[7]

Intense marketing and publicity campaigns accompanied the spread of REA projects and programs. In these texts a contagious enthusiasm seems to propel the new lines into sparsely populated and off-the-grid towns, hamlets, farms, and ranches. As lines approached, farmers purchased distribution wires, light fixtures, and appliances in anticipation of "zero hour"—the moment when the power came on and lights inside their homes flickered to life. That singular moment was a mainstay of REA folklore—families watched electric power and artificial light pulse into their farmhouses and barns. Zero hour was as exciting and historic as a wedding or birth. Children raced from room to room flipping the new switches. Some of the older family members would be brought to tears by the sight of new artificial light bulbs and electric machines inside the same homesteads where past generations had fought tooth and nail for survival. One REA member and convert preached, "The greatest thing on earth is to have the love of God in your heart, and the next greatest thing is to have electricity in your house." The REA transcribed this spiritual testimony for a book title, *The Next Greatest Thing: Fifty Years of Rural Electrification in America* (1984).

The Next Greatest Thing and other REA publications suggest a direct correlation between the new lines in the landscape, the wealth and education inspired by electrification, and the potential success of future generations. Such an association is visually present in an REA poster in which the father in overalls and his son in more modern clothes watch the new lines arriving (fig. 22). A schoolteacher in South Carolina also reported that most students in his area used to quit going to school in the third grade, but "now they go through high school, and many finish college. It all happened after the lines came through."[8]

Some farmers helped the lines "come through" by setting poles, securing conductors, and hanging transformers. Others offered land. As the REA did not originally approve funds to purchase rights-of-way, farmers were encouraged to freely relinquish easements. Between 1936 and 1941 private landowners gave one million separate easements for REA co-ops to build their lines and facilities. Many of these free easements came from families who had already paid their own five-dollar signup fee to finance the new lines. To support the REA with down payments, labor, or access to one's land was to support the local economy and the common good.[9]

Power and the Land (1940), a thirty-minute documentary film, epitomizes the REA's narrative about the transformative power of power lines. The American poet Stephen Vincent Benet wrote and narrated the script. Dutchman Joris Ivens directed the film. Ivens had just finished collaborating

22. Rural Electrification Administration advertisement, ca. 1949. Many REA advertisements suggest electrification creates economic and educational opportunities. Courtesy of the Historic Poster Collections, USDA National Agricultural Library.

with John Dos Passos and Ernest Hemingway on *The Spanish Earth* (1937), a documentary about the Spanish Civil War. Originally, Orson Welles narrated *The Spanish Earth*, but Hemingway felt he could do better, and he narrated a second and more popular version. *The Spanish Earth* opens with peasants digging irrigation ditches in an arid field near Madrid. The fresh waters from the Tajo River will nourish their fields and help them feed besieged Republican forces in the capital. A *New York Times* review said Ivens's film provided "the irrefutable argument of pictorially recorded fact" that the Spanish peasants were not fighting for Marxism or communism "but for the right to the productivity of a land denied them through years of absentee landlordship."[10] Hemingway and Ivens screened *The Spanish Earth* for the Roosevelts at the White House in July 1937. Two years later Ivens began shooting for *Power and the Land*, which, like *The Spanish Earth*, opens with rural folk struggling to make their land productive. In the latter version the Spanish peasants of Fuentidueña are replaced by an American farm family, the Parkinsons of southern Ohio.

In the opening scenes the Parkinsons undertake various labor-intensive chores: milking cows, pumping water, pitching hay, refilling kerosene lamps, scrubbing clothes on a washboard, and ironing by hand. While well-meaning and noble, their work is inefficient. "Good people, hard-working people," Benet narrates, "deserve the best tools man can make." The Spaniards needed irrigated waters to grow food, feed troops, and battle fascism; in the years before that conflict blossomed into the Second World War, Americans needed electric power for health, happiness, and efficiency.

While Mr. Parkinson works with neighbors to harvest a corn crop, the men strike up a casual discussion about electricity. In the next scene an REA representative organizes a schoolhouse meeting. Statistics flash on the screen, and Benet barks out the refrain of "Lights up!" as he boasts of the 500,000 farms, schools, churches, stores, and rural factories enjoying public electricity. Following this triumphant history, the score turns ominous. The screen fills with two slender lattice steel transmission towers in front of a factory spewing smoke. Benet narrates, "The city always had power," and as the camera follows wires from a substation and then across a hilly landscape, his voice rises: "Now wires swing out to the country! They are stretching out, long wires reaching out where wires never went before! There's a tune as the wind blows through the wires . . . Power for the Parkinsons!" Benet's script and his emphatic reading recall Walt Whitman's "Passage to India," which celebrates landscapes with descriptions such as "The seas inlaid with eloquent gentle wires."[11] In Whitman's visions the oceans or landscapes accept the wires; for Benet the wires are actors that swing, stretch, and reach out "where wires never went before."

As a nationalistic documentary, *Power and the Land* harks back to inspiration from the founding fathers. When workers raise a pole into place, Benet explains that, like the first American patriots fighting the British, these men are planting "liberty trees" that will bring freedom and progress to the American farm. After the pole is secured, the camera follows three sets of wires until they arrive at the homestead. In what seems like an overnight transformation, the Parkinsons have power.

In the second half of this before-and-after setup, *Power and the Land* reveals the improvements that electricity has made possible. The Parkinsons have an almost instant familiarity with their iron, hot water heater, stove, and refrigerator. Each of these devices lessens hours of burdensome work for Mrs. Parkinson. Meanwhile, Mr. Parkinson's farmwork is more efficient, and he appears more relaxed because he has a radio announcing weather reports, an "electric mamma" keeping baby chicks warm, and a refrigerator to keep the cows' milk from souring. The new electric technologies are grounded, literally and metaphorically, to community networks. Electrification amplifies rural values of family, hard work, cooperation, and American ingenuity. In the REA campaigns power lines are presented as symbols of freedom, progress, and equality.

In the West the Bonneville Power Administration (BPA) marketed its own massive, government-financed infrastructure projects, including the new transmission lines. In the BPA's first documentary, *Hydro: Power to Make the American Dream Come True* (1939), the narrator describes, "Above the roar of the river and the whirr of the turbines, slender threads of silver swing over the land, carrying the surge of hydro to a waiting people."[12] In 1941 the BPA hired folksinger and rambler Woodie Guthrie as an "information consultant." Guthrie traveled to Portland, Oregon, and signed a one-month contract to write and record songs celebrating the Columbia River, hydroelectricity, and public power. Guthrie says his songs were "played at all sorts and sizes of meetings where people bought bonds to bring the power lines over the fields and hills to their own little places."[13] The importance of building the new power lines to branch out and reach smaller, electrified communities is announced in "End of My Line," in which Guthrie sings: "Coulee dam is a sight to see, Makes this e-lec-a-tric-i-tee, electric lights is mighty fine, If you're hooked on to the power line." Living beyond the lines was reason to sing the blues.

Four of Guthrie's twenty-six "Columbia River Songs" are featured in the documentary *The Columbia: American's Greatest Power Stream* (1949). In "Roll, Columbia, Roll" Guthrie rhymes, "Now there's full three million horses charged with Coulee' 'lectric power / Day and night they'll run the factory and they never will get tired." In "Talking Columbia" he wryly explains that the electricity will run a thousand factories for Uncle Sam to produce plastic "everything." The song finishes with Guthrie naming all the plastic things Uncle Sam needs and then announcing, "Don't like dictators not much myself, / But I think the whole country ought to be run / By e-lec-a-tric-i-te!" As mass electrification and public power converged upon more geographic areas, the social and economic benefits of electricity seemed to outweigh concerns about the "dictators" who may have owned and controlled it.

Together these texts show that electrification is always already social. Ivens's documentary and Guthrie's songs highlight how electrification changed the fabric of labor relations, communication networks, domestic and leisure activities. They also suggest that the sight of average Americans interacting with new technologies and the sounds of witty folky tunes softened and simplified the new, powerful machines sweeping across the landscape.

That transformation, similar to the spread of overhead infrastructure in urban areas, was not always soft and easy. Some rural lines met fierce resistance. Critics refused to accept narratives correlating electricity with progress and government-financed poles and lines as beacons of freedom. As Leah Glaser explains in *Electrifying the Rural American West*, the arrival of modern electric technologies may have accelerated "the disappearance of a distinct landscape and lifestyle between cities and their surrounded hinterlands," but rural farmworkers, ranchers, miners, and Native Americans offered electric power did not automatically adopt urban values or cultural practices. Rather than passively receiving the spread of new technologies and infrastructure, many rural communities "largely determined the need for, access to, and use of electrical power."[14]

The site of lines in the landscape created the sense of agency or lack thereof. Although many rural landowners supported the REA, others felt the cooperatives were trying to take advantage of them. Some farmers living near REA power line routes refused to pay sign-up fees for the co-ops. Also, according to Ronald R. Kline, "large numbers of farm people engaged in a more active form of resistance by refusing to grant rights-of-way (easements) for construction power lines across their land." Obstinate farmers guarded their land with shotguns or waited until the poles were put into place and then chopped them down.[15] In 1938 John Carmody said that organizing four hundred REA projects had taught him that "one of your biggest difficulties will be getting the rights-of-way for your lines. Everybody says he wants electricity, but when it comes to locating the lines and locating poles, many people either refuse to hand out essentials, thereby denying their neighbors electricity, or make it so difficult and so costly that certain lines cannot be built at all."[16] Again, the benefits of electrification, including its attractive icons—lights, home appliances, and farm equipment—could be disassociated from the visible, material infrastructure these devices required.

Power Surge, Aesthetic Scourge

After the Second World War the positive symbolism of rural transmission lines declined. A primary reason was the sheer number of lines required to satisfy the surging demands for power. In 1941 electric power production in the United States had reached 208 gigawatt hours. By 1944 the United States generated 279.5 gigawatt hours, an approximate increase of almost 33 percent in just three years. Total production of electricity dipped slightly during the peacetime transition, but a threshold had been overcome. Electric power production nearly doubled in each of the two decades following the height of the war effort, reaching 544.6 gigawatt hours in 1954 and 1,083.7 gigawatt hours in 1964.[17] Advertising campaigns sponsored by General Electric encouraged returning veterans to "Live Better Electrically" with new appliances such as ovens, dishwashers, clothes dryers, and televisions. Overall, utilities enjoyed widespread public support and relatively lax municipal and governmental regulations. Their shareholders received steady, double-digit returns. Increased production and consumption seemed directly linked to economic growth and prosperity. Everyone seemed to benefit from more electric power.

In the postwar boom period utilities built more massive, cost-efficient generating plants and used high-voltage, long-distance transmission lines and power-sharing agreements to ship "coal-by-wire" across broad regions. Indeed, most of today's balkanized grid and interconnections were planned and built during these years of growth and interstate power-sharing agreements. Before the war lines operating at 235 or 287 kilovolts totaled 2,974 circuit miles; by 1960 there were 22,379 miles of transmission lines operating above 235 kilovolts. Energy regulators predicted that by 1990 the United States would have more than 33,000 miles of 500-kilovolt lines and 159,000 above 235 kilovolts.[18]

At the height of this golden age for the industry, energy policies, corporate agreements, and plans for multimillion-dollar power plants and additional power infrastructure seemed to sweep effortlessly from engineering memos to boardrooms to capitol buildings for approval. However, after electrification became widespread, the warm embrace of transmission lines as signs of progress was challenged by a renewed attachment to pastoral lands and increased sensitivity to environmental damage.[19]

Affluent homeowners and environmentalists made some of the most vocal arguments against high-voltage overhead transmission lines. In the 1940s a utility built a set of lattice steel transmission towers near Taliesin West, the winter home Frank Lloyd Wright had begun building in 1927. Lattice steel towers and a loom of cables sliced through Wright's view of the desert valley and the city of Scottsdale. Some suggest that when Wright learned of the lines that would block his view, he called President Roosevelt to complain and demand that they be moved.[20] However, another source claims Wright made his direct protest to Roosevelt's successor, President Harry Truman. Wright may have complained to both men, but either way it is telling that despite having a valid aesthetic argument and the political clout to have his case heard by the White House, Wright's appeals still failed; to this day lattice steel towers mar the vista from this desert jewel of American architecture. At the time Wright considered moving Taliesin West to Phoenix and starting over; eventually, he reoriented the windows and interior design of the space to draw more focus to the mountains. Not all home or landowners were so fortunate.

After the midpoint of the twentieth century, postwar consolidation of power companies and larger, more centralized power plants meant that more Americans found themselves living near transmission lines. The general acceptance of overhead infrastructure in rural landscapes changed to general uneasiness and then outright disgust. The buzzing cables and skeleton-like structures threatened to electrocute people and animals. The high-voltage current evoked uncertain risks to human health and the broader ecosystem. The necessary land easements infringed upon property rights. Visible lines damaged property values, and the poles and wires constrained irrigation and farming operations. Construction along rights-of-way could harm ecologically sensitive and historically sacred spaces. The utilities argued that the public's list of complaints seemed like a moving target. Even when a new transmission line was planned, financed, and managed to the highest standards, any plans for power lines or semblance of electric infrastructure could stir up opposition. A range of legal arguments and rhetorical devices seemed to calcify the bias against overhead transmission lines. Suburban and rural residents resisted unwanted lines, and some took extreme measures against them.

In 1962 a New York group called "Scenic Hudson" fought a proposal to build a new plant on Storm King Mountain and transmission lines to bring the power into New York City. In late 1969 Consolidated Edison, or Con Edison, tried to compromise by making a series of changes to the proposed power line, including shorter, laminated wooden towers and a route that would avoid ridges and peaks. Scenic Hudson was unmoved: no matter what the design of the line, if the Storm King plant was built, the countryside would be irrevocably scarred.[21] Eventually, after more than a decade of litigation, Con Edison scrapped their plans.

Another public battle erupted the next year, when the Virginia Electric Power Company proposed a 500-kilovolt line across lush wilderness and hunting grounds. After considerable complaints, the utility finally acquiesced to the opponents' demands, moved the line farther north, and incurred an additional cost of four million dollars.[22] A few years later Potomac Edison's plan for a 500-kilovolt power line through Maryland was completely derailed because, according to *Electrical World*, the proposed route "went through too many of the wrong backyards."[23] The implication here is that the "wrong backyards" are those that belong to citizens with the wealth or political capital to complain. The "right backyards" would then be those belonging to anyone who accepts the towers or is unwilling or unable to resist.

Local newspapers sided with the homeowners and also slammed Potomac Edison for proposing that the line cross part of the Civil War battle site of Antietam. *Life* magazine accused Potomac Edison engineers of being zealously devoted to the "holy writ" that demanded transmission lines must follow the cheapest, straightest, and narrowest path. Local zoning boards as well as the U.S. secretary of the interior condemned the line. Critics were outraged to think that if Potomac Edison or another utility wanted to wield weapons of eminent domain, locals would have no power to stop the lines from spilling across their land. Executives from Potomac Edison complained that the opposition was fueled by politics and that instead of posturing for a sound bite, more newspapers and politicians should have listened to their reasons for selecting this route. Sensing an uphill climb for approval and a public relations disaster, Potomac Edison withdrew its proposal.

The industry and its engineers failed to take collective action, and proposals for new lines continued to spark more scuffles. Local newspapers, national magazines, and book-length revelations continued to magnify the real and potential impact of overhead transmission lines. Soon some of the same communities and individuals who had helped set up rural electric cooperatives and welcomed power lines onto their landscapes emerged as some of the most vocal critics of utilities and their infrastructure.

The Northeast blackout of 1965 sparked the construction of more long-distance, high-voltage transmission lines and regional interties. These interties were meant to facilitate an almost instant redistribution of power in the case of a sudden power outage. The new "power pools" allowed utilities to buy and sell power from their neighbors and regional transmission operators.[24] In 1970 the Federal Power Commission stated, "The growing emphasis on environmental protection and aesthetic improvement is to encourage consideration of underground systems but, for the foreseeable future, more than 90 percent of the transmission system expansion outside of highly congested areas is expected to be overhead."[25] The boxy H-frame poles and spiky steel structures that spanned agricultural lands and charged through clear-cut rights-of-way irked more Americans.

In the political landscape the people resisted with public relations campaigns, court cases, and protests, which sometimes turned violent. In the cultural landscape the transmission infrastructure became a flashpoint for various civic debates as the lines touched upon property rights, public health, and environmental regulations. In both arenas hyperbolic rhetoric charged the power line problems and strained the relationships between power and landscape.

Power Line Battles Amplified into Popular Culture

Louise B. Young's *Power over People* (1973) offered the first extended, book-length argument against the negative impacts of new, higher-voltage lines. As an accomplished science writer, Young was at least aware of, if not directly inspired by, Rachel Carson's *Silent Spring* (1962). Carson's landmark case against DDT raised fears about the long-term and mostly hidden environmental and health effects of increasingly complex, decentralized, unregulated industries and chemicals. As part of her argument, Carson noted that electric cable insulation initially included chlorinated naptha-lenes, one of the "elixirs of death" related to DDT. Years of handling the cables caused, according to Carson, "illness and death of workers in electrical industries."[26] What Carson did to raise public awareness about DDT and other chemicals Young wanted to do with high-voltage transmission.

Young's case study was a 765-kilovolt direct current line proposed and eventually built by American Electric Power (AEP) between 1969 and 1972. The route for this exceptionally high-voltage line ran diagonal and ruler straight through southern Ohio, not far from where Ivens had shot his documentary *Power and the Land*. The line would intersect Young's hometown of Laurel, Ohio—population 104. Young depicts Laurel as a village blissfully stuck in a slower, simpler, happier era. Her references to the shattered pastoral, corporate greed, and uncertain scientific impacts relate to the core arguments landowners have used for decades.

First, Young argues that massive utilities and their holding companies attempt to increase profits and market share by inflating supply. Increased supply helps provide cheaper electricity rates, which is enticing for the local communities and the state, but bringing that cheaper power to market requires high-voltage lines to pass through forests, farmland, and small towns. Ultimately, rural Americans must bear the burden of lowering costs for factories and fast-paced city dwellers. Second, Young critiques the energy policies that encourage increasing supply to spark more growth and meet expected increases in demand. If electricity was more abundant and cheap, such thinking goes, then more business and industries will be attracted to a region and an expanding economy will not be threatened by energy shortages. For Young the utility industry's predictions about future demand, lower costs, and potential industrial growth are driven by profits, not efficiency or conservation. New power plants and power lines require a significant financial risk, but if the added electricity supply fails to generate expected revenues, the utility can maintain or raise rates. Utility customers pay for these risky investments with their monthly payments, and rural citizens on the planned route pay for them with their land.

Finally, Young argues that utility engineers and executives decide upon the "preferred route" behind closed doors and that any suggestion that the community can help to select options is not accurate. When the possible routes are announced, right-of-way agents begin visiting homes on the eventual path the power lines will take. These agents sometimes exaggerate how much support the project has received from other neighbors—telling homeowners, for example, that "everybody else has already approved, you do not want to be the last one"—or warning homeowners that certain government or corporate interests have decided they need the line and it is better to sign early, receive a fair payment, and not further complicate the utility's efforts to serve the public. Of course, these tactics, Young suggests, are meant to divide the affected community and limit the information the stakeholders need to effectively organize opposition.

The protagonists in *Power over People* were unable to stop AEP and its 765-kilovolt line. Select landowners rejected the payments offered by AEP right-of-way agents, signed petitions, and published formal complaints in the local paper. Although massive resistance never mounted, Senator Lee Metcalf of Montana advised citizens, to "put [*Power over People*] on your reference shelf next to Rachel Carson's *Silent Spring*" and use it "to obtain protective legislation, better regulatory commissions, increased understanding by the courts."[27] *Power over People* may not be the most effective legal guide for twenty-first-century homeowners, and the arguments about corona discharge and Air Quality Standards or electrochemical oxidants and T V reception are at best outdated. Young's narrative does, however, provide a captivating argument against the environmental impact of power lines and the effects of "progress," which Young places in scare quotes no less than twelve times in her text.

Young's sweeping passages and poignant metaphors portray the ominous "laceration of this complex living system" with visual pollution.[28] The telegraph, telephone, and electric light companies introduced the clamor into American cities and towns in the late nineteenth century. Each of these new networks relied on "lines very similar in appearance," and with each new layer, citizens acquiesced because "five wires were not *much* worse looking that two." Then, after the Second World War, towers began to mushroom over the countryside, and rural Americans became desensitized to the new technologies taking over the pastoral landscape: "We look *around* billboards and *over* superhighways and *under* transmission lines and pretty soon we don't really *see* at all."[29] For Young the American tendency to ignore or misunderstand technological and environmental changes in our midst represents a grave threat.

Power lines can decimate culture, environment, and heritage. Echoing Emerson's essay "Nature," Young says that in the wilderness one can be lost "in the larger organic whole." Conversely, in the "steel forests" one is confronted with metallic scarecrows that rise like the skeletons of skyscrapers and hold heavy lines that lunge in long arcs against the firmament. Since Walt Whitman went forth to celebrate nature, as he explained, "without check with original energy," Americans have been invigorated by wilderness. Yet as the uncultivated spaces disappear, Young relates, "the wasteland is growing every day." Citing ecologist Paul Sears, who wrote in 1962 that "esthetic quality" is the surest and readiest way to diagnose the health of a landscape, Young critiques the apparent sickness of power-lined landscapes. Like Ivens's *Power and the Land*, *Power over People* uses the visual rhetoric of "before-and-after" images. However, Young reaches an opposite conclusion. These black-and-white photos are presented side by side. On the left hand is a nature scene, such as a dense forest with light filtering through the canopy, and on the other is an industrialized landscape, such as a gaggle of poles and wires. Six sets of before-and-after displays are less a clarion call about the utility industry than a warning that the kind of idyllic scenery that once defined the United States had been quickly and irrevocably overcome by industrial blight.[30]

Here it will be helpful to recall that *blight* originally referred to fungoid parasites that spread plant disease or any "baleful influence of atmospheric or invisible origin." *Blight* has also been used as a verb to describe the permanent damage inflicted on something previously bright, beautiful, or seemingly pure (e.g., Lord Byron names "the demon thought" as the "blight of life."). In the twentieth century *blight* was commonly applied to deteriorating urban areas. While perceiving power line blight in urban or rural areas hinges on the visual, Young shows how such perceptions can leak into various cultural and theoretical frameworks. Wire blight spreads ambiguous feelings about place. These feelings can germinate and spread anti-infrastructure sentiment through a community of stakeholders, threatening other technological solutions to environmental damage.

In *Power over People* Young poetically depicts environmental afflictions caused by overhead transmission lines. Minnesota folk hero Paul Wellstone and coauthor Barry Casper's *Powerline: The First Battle of America's Energy War* (1981) argues that arrogant threats of eminent domain made by quasi-public regulatory agencies and utility companies are symptomatic of an unchecked industry and broken energy policies. The economic and political machinations of grid planning also pose broader threats to local autonomy and the preservation of agrarian values and landscapes.

Contrary to most Americans who saw power lines as a fact of life, the CU Project, a 430-mile 400-kilovolt DC transmission line that crossed from central North Dakota to the suburbs of Minneapolis and St. Paul, was "a symbol of America's willingness to sacrifice its rural citizens to feed a gluttonous hunger for energy." For the Minnesota men and women planting, growing, and harvesting food for the American public, the "rhetoric— sincere rhetoric—of saving the good land for food production became a familiar theme of the powerline protest." The urbanites' appetite for electric power, these farmers argued, was trumping the farmers' ability to grow them food.[31]

Casper and Wellstone summarize the labyrinthine procedures and closed-door agreements typically involved in the planning process, but the main feature of their narrative is the transformation of a few dozen conservative citizens into active protesters willing to fight the line by any means necessary. From 1974 to 1976 hundreds of men and women organized to fight the line in local planning meetings and then state courtrooms. When their legal efforts failed, they took to the fields. In addition to a newsletter—*Hold That Line*—and phone calling trees, the opponents used CB radio to mobilize, and whenever a surveyor was spotted in a field, angry and disruptive farmers were not far behind.

From 1976 to 1978, 120 power line protesters in Minnesota were arrested on various charges ranging from trespassing and civil disobedience to assault.[32] One of the first to be arrested was Virgil Fuchs, a resident of Stearns County known for tape-recording every encounter with utility workers and representatives. On June 8, 1976, Fuchs became outraged when he saw company workers on his land without permission. Using his tractor as a battering ram, he destroyed a surveyor's tripod and dented a company pickup truck before the workers fled.

The utilities tried to pacify Fuchs by taking him on a private tour of the Pacific Intertie, a 400-kilovolt DC line operated by the Bonneville Power Administration between Washington and California. The tour was meant to show Fuchs and journalists who accompanied him that a 400-kilovolt line was safe. During the trip Fuchs repeatedly raised questions that representatives could not answer and complained that the tour was rushed and staged like an advertisement for the electric utility industry. When his own protests and legal battles failed, Fuchs said that if he could do it over, he would not have engaged with the courts, the utility representatives, or the press but, rather, would have gone "to one damn public hearing and sent into the public record telling them that you put that line across our farm and I am going to take care of every last one of you guys."[33]

Other locals also felt they should have abandoned civil discourse and legal pathways to focus on more violent and disruptive behaviors. One farmer claimed the utilities were "an evil cartel assaulting individual farm-ers." Another opponent of the same line believed destroying good farmland with power lines was "morally wrong" and "evil." "And all it takes for evil to prevail," he continued, "is when good men and women sit back and do nothing." Combating these evil lines required extreme tactics, as one prominent opponent later reflected: "I don't think it done a bit of good for farmers to go to the courts; I think we could have just as well saved our money and got violent as hell. If you wanted to stop the line, that's the way to stop it. If it ever happens again . . . there's one thing to do and that's get violent as hell."[34] It was unclear where and when to de-escalate the violence. Some protesters, knowing they would be arrested, doused themselves in pig feces to make arrest more unpleasant. Others sprayed ammonia into the air and let it drift toward utility workers and state troopers.

When Alice Tripp was arrested on January 3, 1978, she refused to walk to the police car. Four highway patrolmen picked her up and carried her across a snow-covered field. She later explained: "John [her husband] said he saw me from across the field with my head bobbing along and was quite alarmed . . . It was a symbolic thing. We hoped eventually to get a whole crowd arrested."[35] Tripp's arrest galvanized the resistance and inspired special reports on the CBS and NBC national evening news. With the entire nation watching and no solution in sight, the protests steered toward a dangerous precipice. In March 1978 eight thousand people marched approximately nine miles in subzero temperatures from the town of Lowry to the town of Glenwood. The majority of Americans seemed ready to embrace the farmers' cause. Nine days later a drive-by shooting outside a power line assembly yard in Pope County left a company windshield shattered and a security guard injured. Public opinion turned. After the spring of 1978 the heated assemblies, arrests, and acts of collective civil disobedience waned.

Construction proceeded, but 430 miles was impossible for the utility company to defend. This began an "unprecedented phase of guerrilla war-fare" through the western half of the state. Protesters began by shooting the glass insulators. "Wire worms" (rifle bullets) destroyed fifty-five hundred insulators in less than one year. Then the vandals discovered how to bring the towers down. Operating at night, these "bolt weevils" brought down fourteen transmission towers between 1978 and 1980. Vandalism and security costs added another $6.1 million to the $1.24 billion project. In 1980 the Cooperative Power Association and United Power Association officially requested that the Rural Electricity Association assume ownership of the line so that the local authorities could enlist FBI agents and a federal grand jury to arrest and convict tower vandals of federal crimes.[36]

As the Cooperative Power Association and United Power Association (CU) line neared completion, director Dick Lowry released *OHMS*, a made-for-television movie that, like *Power and the Land*, was shot in rural Ohio. The plot borrows directly from the civil unrest in Minnesota, and the protagonist, Floyd Wing, seems to have been based on the real-life Virgil Fuchs. Early in the film Floyd sees a survey team from the fictitious Unified Power Corporation. The utility intends to shift a line route to cross his land and bring what his son calls "big million volt monsters" onto his farm. Meanwhile, Jack Coker, a long-haired Vietnam vet and local high school teacher, tells his students about "ohms." This is "the unit of resistance to electrical force . . . *resistance*," he repeats slowly and with such inflection that modern viewers might expect him to add *man* or *dude* to the end of his sentence.[37] Initially, the conservative farmers view Coker as another disillusioned liberal ready to protest anything; however, as the battle with Unified Power heats up, Coker becomes a lead organizer for the resistance. Interestingly, like the fictional Coker, Wellstone was a college professor when he became involved with the CU Project. The experience launched him into politics, and he served eleven years as a U.S. senator. In 2000 he considered a presidential campaign. In 2002 he was killed in a plane crash near Eveleth, Minnesota.

The TV promotion for *OHMS* includes clips of an earthmover tipping over a utility truck, a violent brawl between protesters and Unified Power workers, and the image of a lattice steel tower glowing against a sunset while the voice-over exclaims: "Somebody has power! Somebody has none! Somebody is lying! Somebody is going to get hurt! *OHMS*, Wednesday at 9, 8 Central!"[38]

Across the United States artists, architects, environmentalists, scientists, and journalists appeared sympathetic to complaints about utilities scarring landscapes with power lines. Even some executives admitted the industry needed to extend its pre-construction research beyond the technical, financial, and environmental impacts of the line and study the "social human environment."[39] In public the utilities, transmission operators, trade organizations, and politicians expressed a desire for further studies, community engagement, and transparency in the process of planning and building power lines. The electric power industry understood the risk of messy public relations battles and often took a centrist approach to issues of environmentalism and local autonomy. It sympathized with the ideas behind the 1966 National Historic Preservation Act, the 1968 Wild and Scenic Rivers Act, the 1973 Endangered Species Act, and the 1977 Clean Air Act. It could see value in the federal and state-level power plant and transmission line siting laws meant to protect sacred and fragile environments and ecosystems. Indeed, the industry showed a willingness to comply with the various regulations while still working to improve public perceptions of electric infrastructure.

Not everyone in the power industry valued extra input from customers or felt that utilities needed to fund studies of the social human environment. Leaders of the National Rural Electric Cooperative Association felt overregulation and militant environmentalism had needlessly complicated the siting process.[40] Both municipally owned and investor-owned utilities made it their mission to build the most efficient electric infrastructure and to bring cheap, reliable energy to American homes and competitiveness to American businesses. In their view consumer groups and grassroots organizations merely acted as obstructionists and discouraged investment. Those who opposed power lines were unable to recognize that they enhanced public service. Transmission lines in the landscape may be unsightly for some individuals in some places, but if lines had to be moved or buried, costs could skyrocket, and if they were delayed or not built, the community, the state, and even the nation risked energy shortages, system-wide failures, and total social collapse.

Nimrod "Nim" Goldman, the protagonist of Arthur Hailey's novel *Overload* (1979), personifies the pro-energy, neoliberal stance. Nim is the brave, efficacious, and philandering vice president of a massive utility, Golden

State Power & Light (GSP&L), which is a loosely disguised reference to California's largest utility, Pacific Gas & Electric (PG&E). Nim knows the inner workings of the western grid like an engineer and a lineman: he helps uncover an organized crime ring stealing electricity, and he saves a boy that had climbed a transmission tower to retrieve a kite from electrocution and a coal miner from being trapped in a rock crusher. Nim also feels the challenges of keeping the power flowing and simultaneously appeasing state regulators, shareholders, and journalists. The greatest direct danger to Nim, and the entire region, is a small gang of domestic terrorists who sabotage the grid and murder utility employees.

The attacks begin with the bombing of GSP&L's largest power plant, nicknamed "Big Lil." The next day Nim argues that GSP&L must quickly proceed with the Tunipah project, an enormous coal-fired power plant to be built in the California desert. Tunipah, Nim explains, is "North America's answer to Arab oil." The United States has "enough coal west of the Mississippi to supply this country's energy needs for three and half centuries. If we're allowed to use it." Lamentably, even the most logical plans to enhance safety, comfort, and national security are susceptible to capricious contrivances and unreasonable interpretations. A few rascally objectors could stall Tunipah until it was "so long delayed as to be, in reality, defeated." Nim knows this happened during the Storm King controversy in New York. A frantic minority had vociferously opposed the project, and nobody, especially the press, had listened to reasons for cleaner, better, cheaper power. Nim empathizes with the exasperated Con Edison executive who pleaded with the people of New York that at some point "human environment must prevail over fish habitat."[41]

Nim views a significant percentage of Californians as equally unreasonable. At a hearing with the California Public Utilities Commission, a leader of the "Sequoia Club" (i.e., the Sierra Club) testifies that Tunipah would be "sacrilegious, an ecological stride backward to the last century, a blasphemy against God and nature." When Nim takes the stand, an environmental lawyer for the Sequoia Club shows the court photographs depicting the "ugliness" that would be "imposed on what was once a beautiful landscape." To Nim the environmental movement deserves respect, but some of those "who call themselves environmentalists," he says, have their own narrow aims and "what they cannot defeat by reason and argument they obstruct by delay and legalist guile." Eventually, Nim loses his patience with the liberal media, uninformed environmentalists, and "paper-eating bureaucrats."[42] None of them recognize the danger that Nim senses lurking, waiting to take down the grid.

Nim's doomsday prophecies almost materialize because of the Friends of Freedom, a fictional activist group based on the New World Liberation Front (NWLF). In real life the NWLF attacked politicians and corporations across California. Pacific Gas & Electric was a primary target. In 1975 the NWLF planted and exploded pipe bombs at twenty-two sites, including PG&E offices, power plants, transformer banks, substations, and transmission lines. With each bombing the NWLF sent a letter to newspapers demanding free electricity for the unemployed, poor, and retired as well as higher rates for corporations. Surprisingly, such bombings and attacks on electric infrastructure were not uncommon in the United States. Radical militants, terrorists, and extortionists hoping to hold the grid hostage attacked transmission lines in Alabama, Louisiana, New Jersey, Ohio, Oregon, Washington, and Wisconsin. In 1978, the year before Hailey published *Overload*, the FBI reported that a utility or some part of electrical infrastructure in the United States was bombed once every twelve days.[43]

The real-life bombings across the Bay Area targeted various icons of power: banks, stock brokerage firms, and police cars. But the NWLF's most frequent target was transmission towers and substations. In *Overload* the fictional Friends of Freedom attack electric infrastructure because electricity is "capitalism's opiate." Their leader is Davey Birdsong, a university lecturer and mystic who recruits college kids to collect cash donations for environmental causes (which he steals). Birdsong's biggest supporter and accomplice is Georgos Winslow Archambault, a wealthy, Yale-educated anarchist who writes frantic diatribes peppered with Stalin and Gandhi quotes. Together Birdsong and Archambault orchestrate a series of deadly attacks designed to "cut off the electricity, disrupt the core of its system, and . . . thrust a dagger in capitalism's heart!" Their grand scheme is not just to disrupt electric infrastructure but "disrupting—destroying—electricity's people" by planting bombs at the hotel holding an annual convention for the National Electric Institute convention (i.e., the Edison Electric Institute). Birdsong and Archambault expect thousands of civilians to die in the attack and anarchy to ensue.[44]

While some of the bombs explode, a total catastrophe is averted when a journalist, who had previously been critical of Nim and the GSP&L, uncovers the plot and alerts the police. Afterward Nim's estranged wife (he has four extramarital affairs in the course of the narrative), reads the newspaper coverage and admits, "People are beginning to face reality about an electrical crisis." Nim wonders if and when the looting, burning, and plundering that had taken place during the New York City blackout of 1977 will happen in California and possibly spread across the United States.

Overload did not earn much critical acclaim. The *Los Angeles Times* quipped that "Hailey's knack for leaden plots and tin characters is only surpassed by his skill at using clumsy phrases," and the *New York Times* called *Overload* "little more than a thinly fictionalized stockholder's annual report." Meanwhile, the book reached bestseller lists and was later turned into a screenplay.[45] Expectedly, the book received rave reviews from the utility industry. Public relations departments at hundreds of investor-owned utilities received advanced copies of *Overload* and were told the book would "further public understanding of the industry's special problems." San Diego Gas and Electric sent copies of Hailey's novel to state officials as "educational materials."[46] If the movie *OHMS* sensationalized the rural fight against massive utilities and their overhead power lines, *Overload* provided the heroic and fearmongering rebuttal.

In the decades of rural electrification and then exponential postwar growth, a general pattern of responses to overhead electric lines played out across physical and cultural landscapes. After the initial, exciting process of electrification, a community developed an appropriate appreciation for electricity's costs, dangers, and overwhelming usefulness. After a period of adjustment, they began to ignore overhead lines or remove them as far as possible from public view. Slowly, and sometimes subtly, lines in the landscape returned to the field of attention, and individuals questioned how the multilayered infrastructures and organizations controlling the power they required were operated, financed, managed, and regulated. Such questions about the most complex and increasingly necessary machine in human history could rarely be sufficiently answered. This fueled anger, fear, and resentment from the populace. If the choice was between having lines exposed in the landscape or having access to electric power, people chose the lines. Nonetheless, the overhead threads that had once delivered benevolent electricity had been turned into symbols of ugly, brittle, and sometimes unethical blight.

Familiar with this pattern, engineers, utility executives, and industry representatives took considerable efforts to dismiss the characterization of overhead lines as "evil" and the siting and construction of lines as greedy or dishonest. Instead, they worked to generate favorable public opinion, draw attention to the uncertain and costly efforts needed to address the broad range of public complaints, and systematically analyze and predict where, when, why, and how the public might resist their power lines.

Aesthetic Experience and the Battle of Chino Hills

The Tehachapi Project and the conflict between Edison and Chino Hills exemplify some of the competing forces between aesthetics, scientific evidence, public engagement, and construction costs. In this case bureaucratic tools such as the Environmental Impact Report (EIR) failed to help decision makers effectively consider and assess aesthetic complaints. The reasons are twofold. First, perceptions are informed by a combination of rational experience and emotions. The emotional component entwines with aesthetic value. One may look at the line and feel threatened and therefore hate the way it looks. Conversely, one might find something unsightly about its form—the proportions, colors, or angles—and, based on this cognitive assessment, feel disgusted. In short, empirical evidence *and* feelings shape an aesthetic experience. Second, the utilities and their engineers, designers, and architects have had mixed success redesigning towers or hiding them with trees and landscaping. For some individuals, no matter the form or function, calls for anything resembling a power line will stir up resistance.

The utility industry, government agencies, and academics have amassed considerable research on public perceptions of power lines in a collective attempt to limit the negative impacts of the lines and the tensions they emit.[47] In general this research examines public attitudes toward the *process* of planning, announcing, siting, constructing, and managing high-voltage overhead transmission lines (HVOTLS); the *purpose* of the transmission infrastructure and how proposed outcomes will provide economic, political, or social benefits; and the *material impact* of the line on specific environments. Public reaction to real or potential material impacts can act like a series circuit—if one light goes out, they could all go out.

In the series circuit analogy Point A, or the power source, is the plan to build new transmission infrastructure, and Point B, the terminal, is completion of the project. R1 to R4 represents points of resistance. To travel from point A to point B along the circuit path, the proposed transmission line must overcome potential resistance related to perceptions of (R1) land uses and property values, (R2) environmental pollution and damage, (R3) human safety and health, and (R4) aesthetics. Each of these potentially negative impacts can spark public opposition and short the circuit. In this analogy a shorted circuit means the transmission project must be revised or terminated.

Pre-construction public opposition stemming from the first two nodes of resistance are easier to predict, measure, contain, and litigate than concerns about health and aesthetics. The land use, property engagements, and environmental impacts will differ based on geography, route, proposed structures, and the technical requirements of the line, especially its voltage and load. The utility knows the requirements for safe, functional lines as well as the mitigations that should be taken to repair the environmental damage caused by construction activities and the lines that remain upon the landscape. Certainly, utilities and contractors do not always adhere to regulations or follow protocol in securing easements, restoring habitats, and compensating impacted parties. The routing and design process is not perfect, but the industry and the courts have relatively clear standards with which to address public concerns about land uses, property values, and environmental impacts. Concerns about health impacts and aesthetics can be unrulier and more diffuse.

The first nationwide survey of public perceptions, conducted in 1972, suggested that approximately 30 percent of Americans noticed lines during their daily activities and felt a "vague dislike" of all transmission infrastructure.[48] Subsequent surveys on visual salience and negative aesthetic perceptions have shown a broad range of responses: sometimes as few as 13 percent of respondents indicated concerns about the appearance of transmission lines; in others up to 87 percent of respondents thought the lines had a negative effect on their neighborhood's attractiveness.[49] A literature review performed by the International Electric Transmission Perception Project (IETPP) in 1993 found "aesthetics" was cited more frequently than either "health" or "safety" as the most concerning negative impact of transmission lines. Unfortunately, the assorted methods and procedures used to test the aesthetic impact of certain line configurations in various states, countries, and environments made it difficult for the researchers to draw specific conclusions about the relationship between aesthetic perceptions and public opposition. Instead, the researchers called for further investigation of how humans interact with the environment and "variation in the ways that the aesthetic issue has been approached."[50]

Approximately twenty-five years later, the conflict between Southern California Edison and the people of Chino Hills embodied the uncertain influence that public perceptions of health risks and aesthetic impacts can have on a massive transmission project. This power line battle also reveals the shortfalls of public relations campaigns and social science research that attempts to understand, gauge, and mitigate public opposition by linking it to specific demographics or issues. Indeed, an unpredictable array of facts and feelings can shape grassroots resistance to the sight of electric lines on the landscape.

In 2011, after years of legal wrangling, a court injunction paused construction on the line in Chino Hills. As the structures stood dormant and the opposition gathered momentum, Hal T. Nelson and other researchers from the nearby Claremont Graduate University surveyed 358 residents along the proposed route. Thirty-four percent of respondents indicated that their top concern with the Tehachapi Project was the "possibility of health risks" related to electric fields. The next most cited concern was decreased property values (27%) and then visual impacts on the landscape

(11%). The authors concluded that perceived risk, perceived distance (which may differ from geotechnical distance), and the strength of the available communication networks each had a positive correlation with opposition to the Tehachapi Project.[51]

Why did the residents with stronger networks select "possibility of health risks" as their top concern? Between 2009 and 2013 the grassroots organization Hope for the Hills conducted an emotional campaign to show its neighbors the various risks posed by overhead transmission lines and how Edison had misrepresented or downplayed the risks involved. The typical right-of-way width for a 500-kilovolt line is 175 to 200 feet. The right-of-way through Chino Hills was 150 feet. Previous 500-kilovolt lines in the United States had all been strung on lattice steel towers, but due to the narrowness of the right-of-way, Edison used taller, 198-feet steel poles. In the case of a catastrophic earthquake, opponents argued that these poles could crush homes, spark wild fires, or electrocute innocent bystanders. While such a scenario was unlikely, it could not be ruled out. Nor could the potential health impacts of the high voltages. Banners posted to fences and protest signs at rallies informed fellow residents that Edison wanted to use their children "like lab rats." The Hope for the Hills website and "Stop Edison Power Lines in Chino Hills" Facebook page posted numerous links to blog articles about electromagnetic fields (EMFS) and cancer.

Public fears related to the electromagnetic fields emitted by overhead transmission lines might be traced to a 1979 article in the *American Journal of Epidemiology* suggesting children living in homes with high EMFS had a higher chance of developing leukemia.[52] The year 1979 also marked the height of bombings on transmission lines as well as the nuclear meltdown at Three-Mile Island, Pennsylvania. For the next decade legitimate fears and conspiracy theories related to the energy industry gained traction. Journalist Paul Broduer's book-length exposés, *Currents of Death* (1989) and *The Great Power-Line Cover-Up* (1993), made ominous claims about substations, transmissions lines, and faulty indoor wiring. The most damning evidence, Broduer argued, was a 1990 draft report by the Environmental Protection Agency (EPA) titled "Evaluation of the Potential Carcinogenicity of Electromagnetic Fields." The 397-page report, on which DRAFT: DO NOT

QUOTE OR CITE is stamped on almost every page, introduces its findings in large bold font: "While there are epidemiological studies that indicate an association between EM fields or their surrogates and certain types of cancer, other epidemiological studies do not substantiate this association. There are insufficient data to determine whether or not a cause and effect relationship exists." The EPA's Science Advisory Board reviewed the draft but never officially published its findings. The lack of a final approval and the uncertain implications of "insufficient data" inspired conspiracy theorists to conclude that lobbyists, biased scientists, and the Electric Power Research Institute had stopped the EPA from releasing this damning report. Evidence to the contrary was, according to their thinking, unsatisfactory. Numerous large-scale studies and publications by reputable peer-reviewed journals including the *New England Journal of Medicine, Lancet,* and the *American Journal of Epidemiology* have debunked many of the claims about EMFS and cancer. Nevertheless, fears persist, and to supplement the scientific research, the industry has adopted careful guidelines for line voltage and distance from homes.[53]

Epidemiological studies of people living near HVOTL and laboratory tests of EMFS' impact on living cells are ongoing, complex, and contentious. Generally, levels of constant exposure to EMFS below 0.4 microtesla appear to have no observable, deleterious effect on living cells. For most 500-kilovolt lines the levels of radiation directly beneath the line can be 0.5 microtesla, yet within three hundred feet that level drops to 0.1 microtesla.[54] I have taken (admittedly unscientific) readings of EMFS with a field meter and come as close to functional power lines as possible. I have never recorded more than 0.1 microtesla. Still, when I looked at the towers in Chino Hills, I felt tinges of the residents' fear.

In the current discussion of how perceptions of overhead transmission lines inflect the meanings and uses of electricity and landscape, it is neither possible to offer a conclusive statement nor prudent to suggest how best to share scientific evidence with skeptics. For the purpose of my argument it is telling that concerns about health impacts are frequently accompanied by aesthetic judgments. The possibility that EMFS pose an invisible risk seems to either contradict or corroborate the lines' visible blight.

In 1987, near the height of the power-lines-may-cause-cancer panic, ABC *News* anchor Peter Jennings opened a nightly news episode, "We have a report tonight about an unattractive part of the country's landscape that may also be hazardous to your health."[55] The unattractiveness is a given, the added health impacts uncertain. Thirty years later, after hundreds of studies have attempted to abate such fears, health issues continue to be associated with the negative visual impressions. In a 2017 news story about the safety of living near power lines, Regina Santella, a Columbia University professor of environmental health sciences, explained "I probably would not be terribly worried about [living near high-voltage power lines], other than the fact that they are terribly ugly."[56] Such assurances may dispel persistent fears about the carcinogenetic effects of electromagnetic fields, but assuming that everyone agrees the lines are terribly ugly, eyesores, or a blight on the landscape may not allow for those who believe they do not pose a health threat to change their opinion about whether a line should be accepted. If the form is universally unappealing, the function may be more likely to evoke feelings of fear and anxiety.

The perceived health risks are so entwined with the appearance of the lines that when the lines are buried, concerns about EMFs seem to dissipate, even if the magnetic fields do not. If we are conditioned to think of these objects as blight, we may be more likely to also assume they are unhealthy and dangerous. Then again, if everyone agrees they are ugly and nothing can be done to make them more appealing, then those who might bring forward reasonable, evidenced-based health concerns may be too quickly discredited as just being too sensitive to the visual environment.

Defining or debating the aesthetic reactions evoked by infrastructure remains problematic and, some would argue, irresponsible. Why discuss aesthetics when millions of dollars as well as the security, safety, and comfort of an entire region might be at stake? One reason is that the utility industry has not effectively improved the design of their most recognizable objects, and when they have called attention to the lines, the negative connotations have often preempted dialogue with residents. Damage to scenic beauty, visual integrity, and the aesthetics of landscape has been a valid basis for court decisions since the 1960s, which is when parts of the utility industry began efforts to improve the appearance of infrastructure and mitigate opposition based on aesthetic values. Nevertheless, aesthetic attitudes are not easily gauged and debated in the public arena, especially with regards to infrastructure that so few people seem to notice or understand. All stakeholders risk placing too much or too little weight on the thread of aesthetics.

If the transmission structures themselves cannot be improved or made aesthetically appealing, then opposing a line based on aesthetics might lend one to ignore the extreme, possibly unethical financial burden of rerouting or burying the lines. In addition, aesthetic impact can seem frivolous, especially in relation to other security threats (such as terrorism) or environmental quandaries (such as earthquakes, forest fires, climate change, pollution, overpopulation, or loss of biodiversity). If all lines are ugly and they must blight some areas, then opponents can be more easily labeled as NIMBY (Not in My Back Yard), which implies they prefer to hoist the burdens on another community or environment. Recent research suggests that NIMBY labels are often inaccurate, as individual stakeholders who oppose transmission lines often base their positions on attachments to a particular place, desires to protect a habitat, or political beliefs about the lines' owners or the energy sources the lines are designed to transmit.[57] Still, in popular discourse aesthetic objections are often assumed to come from those who live nearby and do not want to look at the lines.

Edison did not explicitly accuse Chino Hills residents of being NIMBY. Early in the battle an Edison spokesperson did explain that burying a small section of line in one specific part of Chino Hills would place unfair burdens on the rest of Californians who desired renewable, affordable electricity. She told one reporter, "I think the protest is interesting . . . 250 households along 3.5-miles want these towers undergrounded, but there are 12 million people in California from Mount Shasta to San Diego who are going to have to split the bill."[58] Two years later the same Edison representative mused, "We don't think 12 million people should have to pay more than $700 million because a few hundred households want a better view."[59] The claims about the cost to the other Edison customers were exaggerated, as the cost of undergrounding, as approved by the CPUC, was

$224 million. In addition to fighting exaggerated claims about the cost of undergrounding, the people living near the line may have not wanted to accept being pigeonholed as merely wanting "a better view." When Nelson and the others conducted their survey with residents, some of the participants may have selected "possibility of health risks" as their top concern to avoid NIMBY labels.

Transmission lines such as the one Edison planned for Chino Hills are more than unsightly or annoying; they are egregious mistakes. Fortunately, in that case the wrong was righted. Continued pressure from the city of Chino Hills and countless hours of organizing and protests from Hope for the Hills turned the tide. The people stopped that line; however, the successful defense did hinge on the lines' visual impact. Lawyers for the city argued that the Environmental Impact Report did not accurately account for the "jarring imprint which the mammoth transmission structures have had on the viewscape." One judge agreed with the claim that the poles' "immitigable and potentially significant adverse visual impacts" posed a particularly high threat to the "scenic integrity and character of the landscape." When the CPUC made its final ruling in 2013, it noted that the "visual, economic, and societal impact" of the tubular steel poles erected in this specific corridor was "far more significant than what the City or the Commission envisioned."[60] To reach this conclusion, representatives of the California Public Utilities Commission needed to more closely scrutinize the Environmental Impact Report; they *also* had to stand next to the tower and imagine what it might feel like to live near it.

The turning point in the battle against the "monsters" in Chino Hills occurred in 2011, when members of the California Public Utilities Commission traveled from their headquarters in San Francisco to stand next to the 198-foot tower in Coral Ridge Park (fig. 23). Two years before the visit, the commission had voted five to nothing in favor of the proposal to place the 500-kilovolt line aboveground on the 120-foot-wide Edison easement. A year after the visit the CPUC voted three to two to force Edison to take down the towers and bury 3.5 miles of the 500-kilovolt line. CPUC president Michael Peevey made the deciding vote and later explained why he changed his decision, a reversal that had a potential

$300 million outcome for Edison and its customers: "I went there and I saw this and I felt it was wrong, that's really all there was to it. I just felt it was wrong . . . [The tower] was not proportional, and it did something that just wasn't right in this community and it moved me."[61] The commissioner certainly took other facts into consideration, but he tellingly cited the visual perception of the lines and the *feelings* they sparked as having influenced his decision.

The Chino Hills power line battle encourages other communities to rethink our current and future power-lined landscapes. First, scientific siting procedures, public engagement campaigns, and environmental impact reports and statements can and will fail to accurately predict or mitigate aesthetic reactions to a transmission line. In the case of the Tehachapi Project, Edison released thousands of pages of maps, diagrams, and schematics. The company attempted to be forward about the potential impacts. The four-volume, six hundred–plus page Environmental Impact Report on the Tehachapi Project released in 2009 addresses how the construction of the substations and new lines could influence freshwater sources, air quality, noise levels, agricultural and recreation land uses, soil composites, wilderness areas, wildlife (including endangered species), and the possibility of encountering native burial grounds during construction. Within this sweeping analysis the section regarding visual impacts on parts of Chino Hills predict a "moderate to high" visual impact on a neighborhood landscape that has "a lack of memorable elements." This was an understatement and, returning to the circuit analogy, a mistake that might be said to have shorted the aboveground option through Chino Hills State Park and forced the line through the City of Hills to be buried underground.

The landscape analyses, drawings, pamphlets, videos, and public engagement materials did not effectively anticipate or respond to how it felt to stand near these poles or to imagine living near them every day. From various angles one was hard-pressed not to imagine the gargantuan poles and thick, growling cables sweeping over the houses like the long wall of netting at a golfing range. If the cables were strung and energized, it seemed as if the landscape beyond the lines would be blocked and the air would be filled with an electric drone. The physical, embodied experience of standing

23. Steel tower for a 500kV line erected by Southern California Edison in Chino Hills, one of the 500kV towers in Chino Hills to have been approved in 2009 but then deemed unacceptable by the California Public Utilities Commission and removed in 2013. Photo by Jan Palmer, 2012.

nearby the right-of-way and imagining the negative impacts of the finished product shaped the public reactions and the final court decision.

The second insight from the Chino Hills power line battle relates to multimodal rhetoric. Our ability to compose and share image, text, and video allows us to shape how public infrastructure appears on the local landscape. Systems built with the most advanced engineering designs and cutting-edge scientific research are still subject to external forces, including how people see, write, share, and think about them. This is especially true in the age of social media and the ability for a single comment or reaction to a proposed project to go viral.

During the conflict in Chino Hills, the city webpage provided official news and updates related to the ongoing court proceedings and construction activities. Meanwhile, Hope for the Hills created an impressive web page, numerous videos on YouTube, and unforgettable neon T-shirts, pictures of which were posted across Twitter and Facebook. Hope for the Hills activists also sent floods of letters, postcards, and faxes to Edison headquarters and the homes of its executives. They reached mainstream media outlets and their neighbors by elevating the message and welcoming others to join their nonviolent and persistent campaign. President Bob Goodwin told me years later: "We made the line personal. We showed the public the faces of the families that would be impacted."[62]

Over the course of a few years a group that had gathered to fight a power line began to form strong community bonds and practice intense civic engagement. The resistance was not merely YouTubed or Facebooked. Joanne Genis, another leader of Hope for the Hills, had no experience with legal, governmental, or regulatory procedures until she received a pamphlet from Edison in 2009. Joanne and I shared emails throughout the conflict, and she often included attachments with the legal briefs and passages from the California Environmental Quality Act. When we reconvened after the battle had been won, in 2015, she cited rules and regulations that she felt that the California Public Utilities Commission and the utility industry needed to revisit and adjust.

For their successful defense of the city, Coral Ridge Park was aptly renamed "Hope for the Hills Park." Group members still gather each year to celebrate, but Goodwin also admitted that neither side had earned a clean victory. The towers needed to be removed, but they never should have been put there in the first place, and the underground line still makes some people anxious. "Edison never really took into account the community values," Goodwin told me. "They do environmental impact reports, but not human impact reports or family impact reports. I hope they learned their lesson. We were not going to go away easily."[63] May the shared experience of fighting against that power line continue to be directed into demands for more environmentally friendly, energy-efficient, and sustainable community bonds.

For almost a century stakeholders on either side of the line have worked to mold the force of public opinion to suit their own ends. Attitudes and arguments tied to property values, safety, land use, EMFS, corporate consolidation, government control, ecosystems, wildlife, and eminent domain have threatened to derail multimillion-dollar (and sometimes multibillion-dollar) construction projects. During the second half the twentieth century and first part of the twentieth, utopian visions of America's industrial ascendancy have given way to greater technological pessimism and a dawning environmental consciousness.

The narratives show that high-voltage overhead transmission lines brim with tension and resistance. Engineers, executives, architects, and construction companies must balance economic realities, energy forecasts, and political operations with complex equipment and the laws of physics. Power engineering plans, geographic surveys, and real estate analysis must gel with specifications for cables, insulators, switches, transformers, and substations. Diagrams and charts predict the conductivity and resistance of metallic cables and how they will constrict in cold weather, sway in the wind, and expand and sag in the heat. Ground clearances and distances between the dispersed cables and their glass insulators are measured to the centimeter to prevent potentially lethal discharges. Even after the line is energized, engineers, system operators, line workers, and arborists must continually monitor and maintain the lines to keep the currents synced and system load balanced.

Power lines also emit cultural tensions. The poles, towers, and dipping lines represent the ambiguous effects of industrialization and the shifting experiences and values of technology and the American landscape. Sometimes the lines are accepted for the positive economic or environmental benefits they represent. Often the real or imagined views of tall, wide, or high-voltage lines spark public resistance. Outcomes vary, but when those fighting the line succeed, the line's route or design tends to be significantly altered and thus require fresh rounds of reviews, permits, and approvals. For the utility to alter its plans means extra financial costs, which can be passed on to shareholders, ratepayers, or, for some federal and public power projects, taxpayers in a distant city, county, or state. Sometimes the utility cuts its losses at this juncture and the proposed line is canceled.

The tensions emitted by power lines remain, but an energy transition is under way. That transition demands answers to difficult questions: Which groups or organizations are responsible for mitigating a lines' negative impacts or making them less visible? As we move to a more renewable grid, what forms and functions does the public expect to electrify our lives and our environments? What does a sustainable infrastructure look like in the landscape, and who will build it?

Continuing to evaluate the rhetoric surrounding the siting and placement of overhead transmission lines can clarify the cultural impact of electric transmission and contribute to ongoing conversations about how new infrastructure should relate with the landscape. Within the power-lined place, infrastructures and technologies are subject to internal changes (such as voltage surges), external forces (such as weather), and the social impact of public perceptions. Beauty and blight may stem from the eye of the beholder, but the poetics of infrastructure reveals some of the deep fissures between the public and its power companies and between the modern need for electricity and the heritage of the American landscape.

Conclusion

The preceding chapters traced forms and functions of electricity and American landscape, and called attention to the wires where these two social constructs coalesced. The proliferation of overhead, long-distance electric infrastructure was marked by technical leaps and culture shocks. The pervasive lines used to keep distant markets and communities *in touch* interrupted the individual's access to the energy stored in the magnetic, wild, *untouched* landscape. Some Americans read telegraph and power lines as proof that their city or town was "on the line" and visibly powerful; generations later, pushing the lines out of sight or getting "off the grid" would be increasingly difficult. Electric lines repeatedly unsettled idealized visions of landscape, including the wilderness, cityscape, the suburban neighborhood, and the family farm. Representations of overhead lines in paintings, poems, novels, and films endorsed and simultaneously challenged the popular coupling of electricity and progress. As electric networks imposed a new order on the land, an increasing reliance upon, and addiction to, electricity and electric technologies contributed to an expanding, ubiquitous wire-scape.

Our grid is a massively complex and technologically advanced marvel, but the financial, bureaucratic, and social traditions that intersect it limit our ability to rewire energy practices. Efforts to build a better grid will require interdisciplinary analysis of electric transmission's diverse impacts, including costs, environmental damage, and aesthetic practice. The way toward a smarter, more equitable, and more aesthetically satisfying experience of transmission lines is not to turn the design or management of energy systems over to artists and academics. However, the solution, or solutions, is also not as simple as favoring engineering or aesthetics, renewable energy or nature, progress or heritage, economics or environment. Rather, the problem is the entangled infrastructural and emotional forces that undergird electricity *and* American landscape.

The Power Line Piece of the Infrastructure Puzzle

Citizens of developed nations overlook the more banal parts of their infrastructure, such as pipes, wires, and undersea cables. In a seminal chapter on infrastructure and modernity, Paul Edwards argues that "mature technological systems—cars, roads, municipal water supplies, sewers, telephones, railroads, weather forecasting, buildings, even computers in the majority of their uses—reside in a naturalized background, as ordinary and unremarkable to us as trees, daylight, and dirt."[1] While I contend that transmission and distribution lines are almost as remarkable as trees and dirt, it is also clear that in recent years energy infrastructure has drawn more consideration—often because of its failures.

Since 2001 the American Society of Engineers has given the nation's electricity grid a grade of D or D+. While for most of us, the grid is powerful and reliable, others view it as an inefficient machine. One study suggests Americans suffer from more power outages than any other developed nation.[2] Another points out that part of the network loses power 285 percent more often than it did in 1984. Significant outages are also on the rise: 15 in 2001, 78 in 2007, and 307 in 2011.[3] These hours- and days-long outages further hinder productivity and require more extensive repairs. Losing power for even a few minutes can damage sensitive machinery and electrical components. Overall the total cost of power outages on the economy is estimated to be between $80 and $188 billion a year.[4]

Some outages are the result of outdated transformers and power lines. Approximately 70 percent of the nation's transformers and 70 percent of its high-voltage transmission lines are more than twenty-five years old. Most of this equipment has a thirty-year lifespan, and significant rebuilds are under way. At the same time, the costs of outdated and inefficient infrastructure are mounting, and engineers, energy experts, and scholars have been sounding the alarm.

Most customers only notice infrastructure when it fails. For as long as wires have been strung overhead, the forces of nature have intermittently brought them down. Between 2003 and 2012, 679 severe weather events, including record-setting blizzards and hurricanes, downed thousands of miles of transmission and distribution lines and caused considerable deaths and other losses. In 2017 utilities reported seventy-two severe weather outages that affected at least fifty thousand customers for an hour or more.[5] The Department of Energy predicts that the average cost of weather-related power outages is between $18 and $33 billion every year. With the recent string of devastating hurricanes, the annual costs of outages seem poised to cross the high-water mark. Multiday outages have crippled power systems around the Gulf of Mexico and throughout the southern United States. After hurricane Maria, Puerto Rico was without power for months, and it could take years before power on the island is fully restored.[6]

Severe weather and "acts of God" wreak focused havoc, but incremental climate change also strains and damages the grid. Faulty or fallen transmission lines have sparked a number of wildfires in the West. The *LA Times* reports that power lines have been the leading cause of California wildfires in recent years.[7] Droughts threaten hydroelectric and thermal generating plants that require freshwater. Heat waves spike the use of air-conditioning, adding to the risk of overloads.[8] As the earth warms and the intensity and frequency of extreme weather increases, many of the lines we currently rely upon will be susceptible to storms, flooding, fire, and drought.

Ironically, with its reliance on coal, the energy industry has been a primary culprit and victim of climate change. In 2016, 17.5 percent of all residential electricity use was spent on space cooling and 9.1 percent on space heating. That year the U.S. electric power industry produced 1.925 million metric tons of carbon dioxide, or 39 percent of all energy-related carbon emissions. Of these, coal produced 1.364 million metric tons, or 71 percent.[9] Although hydroelectric dams on Niagara Falls, across the West, and throughout the southeastern United States shaped the first decades of electrification, coal provided the bedrock of the modern power grid.

This high-energy, transportable, and storable fossil fuel has been a more reliable energy source than falling water, wind, or sunshine. Coal can be mined, shipped, and stockpiled near a power plant; it offers kilowatts on demand. Without coal the U.S. power industry and the nation's status as a global power might not have been possible.

The market systems and social practices built around cheap, readily available coal helped to divorce electricity consumption from production. Consumers did not see the coal burned to generate electricity heated, cooled, and lit expanding metropolises. Instead, they saw the long-distance transmission lines string out the suburbs. The use of coal in the electric utility industry peaked in 2010. It may never recover, as the cost of natural gas and large-scale renewable has dropped considerably.

In addition to the move away from coal, four other forces are encouraging grid upgrades and energy transitions: sustainability, competitiveness, integration, and security. Concerning sustainability, it seems clear that expanding economies and rising populations cannot permanently scrape and burn the earth's coal dead bones and create radioactive waste for the next century without catastrophic effects. Engineers and world leaders are actively seeking long-term and sovereign solutions to their energy needs. Meanwhile, renewable generation is becoming cheaper and more competitive. Large solar plants and wind farms are appearing with more frequency on the landscape and the seascape. The upfront costs for renewables are falling, forcing the retirement of older, less cost-effective generating plants (especially coal and nuclear).

Deep-rooted coal infrastructure and the efficiency and low cost of natural gas suggests that the United States will not abandon fossil fuels anytime soon. It should also be noted that solar panels, wind turbines, and other forms of renewable, or "clean," energy have their own dirty secrets.[10] In 2014 a report in *Nature* revealed that in China, one of the global leaders in renewable energy output, "production of polysilicon and silicon wafers for solar panels creates dangerous by-products, in particular silicon tetrachloride and hydrofluoric acid, which are being discharged into the environment after inadequate waste treatment."[11] The copper mining practices that support renewable energy also inflict significant environmental damage. Conservation is critical; we can individually and collectively consume less energy and produce less waste. At the same time, a thicker renewable portfolio and a less centralized, more distributed bulk transmission system should also speed the retirement of the greatest energy polluters.

Bringing cleaner energy sources online will require a different grid. For starters the future grid must be more adaptable. This means sending and receiving variable loads from multiple points: traditional power plants, large-scale renewables, smaller community wind farms, rooftop solar panels, home battery systems, and electric vehicles. The so-called smart grid will utilize millions of software sensors, microprocessors, and automation devices attached to switches, circuit breakers, and bus bars. The information gathered across the grid will "allow transmission lines to communicate with each other."[12] The smart grid promises to decrease transmission losses and provide network operators the ability to more quickly and efficiently shift loads based on weather, demand, and other unforeseen fluctuations. The next generation of power networks will also have more capability to instantly adjust thousands of connected thermostats one or two degrees at various times of day to reduce the need for standby power plants and the risk of overloads. Research suggests individuals and communities will be leery of letting a utility control when and how they use electricity.[13] If thousands of customers agree to periods of slight discomfort and the grid can more heavily rely on large-scale renewable power, utilities could decommission some the standby plants that they maintain only for the warmest days, when demand for air-conditioning spikes.

A final piece of the infrastructure puzzle is security. This is not a new concern. Morse worried that vandals might destroy his first telegraph line if he strung it aboveground. In the late nineteenth century sneaky wiretappers, so-called overhead guerrillas, and wire sabotage in the West laid bare the vulnerabilities of overhead lines in crowded metropolises and unsettled frontiers. A surprising number of bombings and other small attacks on outdoor transmission towers and substations in the 1970s caused numerous blackouts, especially in California. In 1982 Amory and L. Hunter Lovins published *Brittle Power*, a report that considered how a coordinated attack might spark blackouts across the continent. They concluded that attacking energy facilities was arguably the cheapest and most effective way for military forces and terrorists to inflict widespread damage and chaos.[14] While not often publicized, the grid and those tasked with protecting it face significant threats. In 2017 U.S. utilities reported twenty-four cases of sabotage and vandalism to the Department of Energy; five of these acts disrupted service for 30,766 customers.[15] The damage could be much worse. In 2012 Defense Secretary Leon Panetta warned that a terror attack on the transmission grid could either cause or accompany "the next Pearl Harbor" for the United States.[16] As the grid goes digital, engineers worry that a virus or act of cyber warfare could potentially trigger cascading failures. If so, restoring power would mean containing and neutralizing the threat and rebuilding sections damaged by severe swings in voltage. Equally unsettling, in 2018 the Pentagon announced that a widespread cyber attack on American infrastructure could merit the use of the nation's nuclear arsenal.

Forecasting energy policies or predicting how the grid will function in 2030 or 2050 seems futile. Politics, fluctuating fuel prices, climate change, war, terrorism, self-driving electric vehicles, massive storage batteries, and killer apps connected to the internet of things promise that our relationship to energy will significantly change in ways that are difficult for energy experts to anticipate. Ideally, the future grid will be more secure, more adaptable, and will offer communities and individual consumers more decisions about what type of energy—coal, natural gas, solar, wind, hydro, geothermal—puts power into the grid and how we take it out. What does seem certain, for better or worse, is that the grid will require overhead wires to transmit bulk power.

Investment in transmission infrastructure has risen steadily over the past decade. In 2016 capital expenditures reached $21 billion. The Edison Electric Institute forecast that transmission investment would peak at $22.5 billion in 2017 and then taper off slightly.[17] Further transmission investments could make the grid more sustainable, more secure, and more efficient. In 2016 Alex MacDonald and colleagues at the National Oceanic Atmospheric Administration built a model to show how a nationwide network of high-voltage direct current lines operating near 765 kilovolts could meet current and future U.S. energy demands. In their model this vast network would transmit energy from giant wind farms and solar fields located in central regions of the continent to the more populated and energy demanding coasts. Although a 100 percent renewable system may not be possible and even an 80 percent renewable portfolio may suffer from interruptions on cloudless or windless days, being able to ship renewable power from any point in the country would take advantage of economies of scale and limit the risk of running out of power. The research team concluded that a sustainable, secure, nationwide electric power grid could satisfy national demand and reduce carbon emissions by 80 percent from 1990 levels.[18]

In terms of epic infrastructure projects, this high-voltage, nation-spanning system would be on par with the interstate highways built in the 1950s. Such a system would reap significant economic and environmental benefits as well as make renewable energy a more permanent part of our lives. It would lessen transmission losses and mitigate the potential havoc caused by failures, storms, and attacks. The investment would likely pay dividends. While the United States has the scientific, technological, and engineering expertise to build such a massive system, other constraints may be too great. As one energy expert explained, "The problem is not rooted in technology, but rather in the way that the U.S. power system is organized legally, politically, economically, and culturally."[19]

The gridlock is both a cause and result of the balkanized system. In 2018 the Federal Energy Regulatory Commission, the National Energy Reliability Council, and 51 other state and city commission oversaw the development and operation of U.S. energy infrastructure. Such oversight may seem necessary, considering that in the same period 192 investor-owned entities controlled 80 percent of the entire transmission infrastructure in the United States. By comparison, however, 76 percent of Americans received their broadband Internet service from just four corporations—Comcast, Charter, AT&T, and Verizon.[20] The sheer number of electricity-generating sources is astounding. At the start of 2017 the United States had almost as many significant generating facilities (8,084) as it did Starbucks stores (8,222).[21] Not all of these are major plants, but they can each produce at least 1 megawatt, or approximately enough electricity to power a thousand homes. They connect to 600,000 circuit miles of transmission lines, 240,000 of which operate above 230 kilovolts.[22]

The vast web of transmission lines that stretch across the North American continent are divided into three interconnections: Eastern, Western, and Texas (fig. 24). The Eastern connection is further divided into eight subregions from Florida to Ontario. System operators, power pools, and reliability councils work within and sometimes across regions. The operators, pools, and councils can have distinct specifications and market designs. Transmitting power from neighboring states, such as from Utah to Colorado, is a challenge. Many regions seem woefully splintered. Recalling Thoreau's critique of the telegraph, electric messages and live video-streams can be sent and received all the way from Maine to Texas in milliseconds; however, it is difficult to send bulk power from Maine to New England.

The Renewable Transition Meets the Urge to Go Underground

Stakeholder engagement, public perceptions, and aesthetic impacts, far from trivial details, remain significant barriers to widespread reform of the U.S. power grid. If federal government and the utilities agreed upon a multitrillion dollar plan to modernize the grid, developers would still "face opposition from landowners who would not want their property bisected or their views obstructed by unsightly power lines."[23] Individuals and communities may want to invest in renewable energies and link to a transcontinental grid, but they are also likely to resist modifications that require *seeing* more overhead transmission infrastructure.

For many stakeholders adjusting their own aesthetic attitudes or finding ways to reframe the lines in the visible landscape is not as attractive as a simple mandate: put the lines underground. Underground lines, many customers assume, would be safer from storms and attacks. It would seem to eliminate the noise made by the lines and the potential health risks of living near them. Underground lines also do not seem to blight the landscape. Anticipating the urge to take lines underground, the Edison Electric Institute regularly publishes a report titled *Out of Sight, Out of Mind*. The title stresses the overarching dichotomy the utility industry

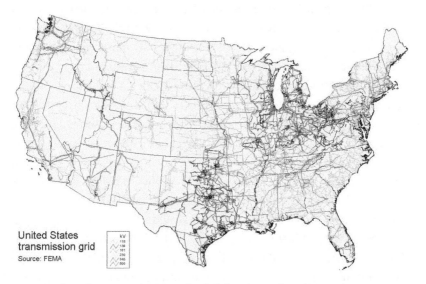

United States
transmission grid
Source: FEMA

kV
115
138
161
230
345
500

24. Map of North American Regional Reliability Councils and Interconnections, ca. 2011. Wikimedia Commons. Created from data by the Federal Emergency Management Agency via the National Renewable Energy Laboratory.

wants to maintain between *sight* and *mind*. The title seems resigned to accept the lines' aesthetic limitations. Sure, no one likes to look at the lines, but no matter how unsightly the lines may appear, keeping them within sight means also keeping them in mind, which means paying attention to environmental impacts and efficiency. The report weighs fairly the various costs and benefits and incorporates externalities about specific lines and environments into its calculations. Underground lines may be less likely to fail during wind-, rain-, or snowstorms, but finding and fixing problems with lines underground is more difficult and time-consuming, which might in turn lead to longer outages than if the lines were overhead and could be quickly diagnosed and fixed. Underground lines are also not secure from floods and earthquakes. *Out of Sight, Out of Mind* reconfirms the role of aesthetics, explaining that the primary driver of the desire to move lines underground will remain "improved aesthetics and the hope that underground electrical facilities will provide greatly enhanced electric reliability."[24] Again, it is not an accident that "improved aesthetics" is linked to the "hope" of reliability. Aesthetic disgust with the lines is valid, yet from the industry's standpoint, remedying it may not improve reliability enough to bear the costs of putting the lines underground.

The uncertain impact of aesthetic improvements and extreme cost of underground solutions drives the industry's preferred dichotomy of *sight* and *mind*. That dichotomy is not necessarily made in the public interest; the utility and its organizations make their arguments based heavily on maximum efficiency and profits. Nevertheless, each severe storm and power outage inspires the media to take up the underground issue. The headlines follow a familiar pattern: "If Power Lines Fall, Why Don't They Go Underground?" (NPR, 2012), "What Would It Cost for the U.S. to Bury Its Power Lines?" (*Fortune*, 2017), and "Isn't It Better to Just Bury Power Lines?" (CNN, 2017).[25] Like the industry report, these articles question the costs and benefits, the logistical tangles and potential financial burdens. Building new underground lines cost six to ten times more than overhead, requiring $500,000 to $2 million *per mile*, depending on voltage, geography, and population density. Communities therefore must consider whether putting lines out of sight is worth incurring a massive debt.

If customers will resist more overhead lines and will not pay inflated electricity bills to take them underground, what might make overhead lines more acceptable? A 2009 survey by Saint Consulting Group observed that support for new lines rose from 46 to 83 percent when "respondents are asked specifically about high-voltage transmission lines delivering wind power."[26] Other anecdotal cases corroborate these findings. In 2003 Xcel Energy announced plans to build a transmission line through rural Minnesota to carry renewable wind energy to market. Rather than resist the new lines, some farmers wanted the lines or substations moved nearer to their land. They hoped to decrease the interconnection costs for their wind turbines. For this group the new transmission line promised to increase their personal profit margins.[27] Xcel Energy touted this finding as a signal that renewables could change attitudes about transmission lines, but without a direct incentive, most home and landowners will continue to resist.

Numerous studies suggest that opposition to new power grids varies.[28]

Still, public opposition seems statistically balanced across North America and Europe. In one of the largest surveys of its kind, residents from twenty-seven European countries were asked how they would react to a new overhead transmission line and were given different reasons for why the line was necessary. Even if the proposed line would deliver renewable energy or provide economic advantages, 34 percent of respondents indicated they would "definitely not accept the new project without opposition." When tested against the alternative scenarios, the researchers concluded that ancillary information about specific environmental or economic benefits of a line could improve public acceptance by a few percentage points, but the percentage of respondents likely to select "definitely not accept" or "probably not accept" would still be significant.[29]

In 2016 I conducted a study using the same scenarios used in the European study. Although 54 percent of the (admittedly small) sample of eighty-two participants in Omaha, Nebraska, indicated that overhead lines "do not bother me," 32 percent of the respondents said they would definitely oppose a new transmission line if it was sited near their home, and 33 percent said they would probably oppose it.[30] The existing lines in a metropolitan or suburban area may be ignored; the idea of new lines that deliver environmental or economic benefits may be welcomed; nevertheless, overhead lines will remain anathema to a substantial part of the population.

A general law of thirds seems to apply to transmission projects: one-third of the public will definitely oppose any nearby, visible transmission line regardless of its potential to lower costs, improve security, or decarbonize; one-third will approve of it (or not care); and one-third will be inclined to oppose it but may be swayed to one side or another by arguments and facts showing the line's potential benefits or setbacks.

How do utilities and transmission companies meaningfully engage the public with new infrastructure projects? A 2015 report from the U.S. Department of Energy on the need for new transmission infrastructure said developers must "engage the public early" and respond meaningfully to concerns to "pre-empt or at least mitigate the impacts of some forms of organized opposition."[31] Early engagement that preempts opposition is not the same as meaningful dialogue. Transmission line owners and operators are often accused of adopting "the rhetoric of deliberative engagement whilst lacking a clear rationale and effective means to incorporate citizen perspectives in long-term network development or specific infrastructure siting proposals."[32] Utilities mail out notices, encourage feedback, and hold public forums, but a significant portion of the public believes that the engineers, executives, regulators, lawyers, and bureaucrats dictate where the line can or cannot go and then use outreach teams or litigation to justify their route.

Power-lined landscapes will continue to be contested spaces framed by incongruous and sometimes conflicting powers. Politicians and community leaders may campaign for "shovel-ready" construction projects and more jobs, but it would be rare to hear a candidate announce an intention to bring more overhead powers lines to her or his constituents. If utilities or government entities fail to effectively address the issue or rely too heavily on the threat of eminent domain, the public will continue to resist change. In the ensuing struggles about transmission routing and design, facts will be culled, packaged, and delivered by utilities and environmentalists, industry associations and consumer groups, and a broad range of federal agencies and nonprofit organizations. Communities and homeowners will receive policy briefs, media articles, and enhanced images and diagrams juxtaposing the new poles and towers with repurposed transmission corridors and rooftop solar installations. Some outreach teams, designers, and engineers will continue to hide their lines, but others could highlight sleek tower designs and even brand their lines with corporate logos. If Apple, Verizon, Amazon, or Google managed the smart grid of the future, would they stamp the lines with their icons?

Whatever future forms and functions the grid adopts, upgrading the most expansive and expensive machine in human history will require an astonishing coordination of science, engineering, finance, and politics. These herculean efforts will need public support. As I mentioned in the beginning of this book, grid literacy is crucial to rewiring the material grid and the social structures embedded within it. As customers and global citizens, we need facts and narratives that apply a balanced mind-set to our many energy needs and problems. Books, reports, infographics, news articles, lectures, and videos can reveal some of the layered operations and far-reaching consequences of charging our phones, adjusting our thermostats, or streaming our favorite videos. However, promoters of the newer, more renewable grid must also address the power line problem.

Fresh angles and new aesthetics can alter ways we see the lines running into our offices, classrooms, libraries, and homes. Understanding voltage drops, kilowatt-hours, power-sharing agreements, and even "shadow prices" (the estimated price for electricity at a point in the power network at a specific time) can demystify the complex and expansive webs that carry current through our walls and above our heads. This enhanced literacy may also inform more creative texts and activities that illicit wonder and excitement for previously unconsidered infrastructure.

Playing a Poetics of Power Lines

Humanities scholars and social scientists are offering novel approaches to energy issues. For instance, Benjamin Sovacool says that to reveal the myths regarding the ephemeral energy systems we all use and predominately overlook is akin to "investigating the invisibility of an already seemingly invisible set of technologies."[33] As long as the lines in the landscape congeal in the background, it is easier to accept myths of wire evil, cancer clusters, and corporate maleficence. Sovacool is one of a growing band of scholars applying a range of social science and humanities methods to electricity systems and practices. The results—conferences, peer reviewed articles, websites, art installations—are painting a more interdisciplinary picture of infrastructure's role in our current and future energy landscapes.[34]

The turn to infrastructure has also captured the attention of various disciplines: literary theory, digital humanities, and anthropology.[35] Paul Edwards clarifies some of the ways scholars might read infrastructure, explaining that these "artificial environments" "simultaneously constitute our experience of the natural environment, as commodity, object of romantic/pastoralist emotions and aesthetic sensibilities, or occasional impediment."[36] Related to these efforts to rethink infrastructure, anthropologist Brian Larkin has suggested that a poetics of infrastructure be applied to social objects such as wires, pipes, roads, and internet protocols. He sees infrastructures as material, circulatory systems that transmit people and goods as well as networks of referents. Larkin imagines infrastructure operating as "concrete semiotic and aesthetic vehicles." Perceptions of the infrastructure's principle forms—or its poetics—are related to their political force, but unique or quirky design features and other "fetish-like aspects" of infrastructures remind us that these systems can circulate meanings "autonomous from their technical function."[37] The millions of tools and materials embedded in our landscapes—glass, steel, aluminum, plastic, or organic light-emitting diodes—attract and shape what it means to be human in a particular age. Common infrastructures can gain a special attraction and symbolic meaning.

To support Larkin's argument and fold it back into the humanities, I want to conclude by offering two approaches to a poetics of power lines. Calling for a poetics is not a plea for aesthetics alone. Of course, tower and pylon designs may improve, and landscape architects, industrial designers, and utility engineers should follow the general guidelines with regards to rights-of-way, tower designs, and landscape modification that have been advised since at least the 1970s.[38] When possible, limit the use of lattice steel towers, route major lines through existing corridors and industrial areas, keep the taller pivots away from areas of high visibility and high visual quality. And when possible, feather the corridor with trees or shrubbery. The wire-scape should not dominate a landscape.

A poetics is a way of teasing out the lines' layered meanings, learning to read the lines' varied impacts, and accepting electricity as an emotional and visual element in what is, ideally, a balanced pattern. The poetics I am calling for has two branches: one works to understand the lines as actors that intersect and create space; the other encourages viewers to pause and gaze upon these somewhat weird wire networks as forms that occupy "place."

The first has been, and may continue to be, achieved through lines of poetry, descriptions of scenery, video clips, and any other genres and mediums that naturally engage movement through landscape. Since the age of the telegraph, individuals have seen electric lines and imagined them radiating, spreading, or marching across the horizon. The lines seem animated with a smooth, continuous movement. In this mode the moving lines are like projections of thought. The wispy wires and the metallic spires might be repeated like a variable line of code—the mimicked appendages, the filmstrip's cells, the patterned tiles, the bar lines and repeat signs on a music sheet. The poetics of power lines desires more than metaphors; the physical manifestation, siting, and construction must insert the code into the operating systems; the poetic and technical aspects may converge. For instance, consider the code words for the industry-standard bare aluminum 1350 conductors and the aluminum conductor steel-reinforced (ACSR) cables. The .12-inch aluminum cable with thirty-seven wires is "mistletoe," and the .31-inch ACSR conductor with 6/1 stranding is "sparrow." Others aluminum cables are nicknamed "peony," "daffodil," and larkspur"; other ACSR names include "raven," "pelican," and "cardinal."[39] With a spotter's guide or app to identify the wires and match them to their bird and flower names, one might read the lines passing by as creative symbols of the organic and rhythmic landscape. Peony . . . cardinal . . . larkspur . . . jump!

In addition to identification and code work, a poetics of power line movements might also aspire to project the technological sublime. The great engineering achievements of the last two hundred years have been shocking revelations of form and power. The thunderous railroad weaves through the pastoral scenery; the George Washington Bridge leaps from city to the cross-country artery I-80; the Golden Gate, three thousand miles away, presses toward the great blue Pacific; the twenty-story space shuttle launches and then seems toylike against the grandeur of outer space. To match these great achievements and spectacular displays, the function *and* form of power lines must find proportion, suitability, and order. This power infrastructure might illicit interest about the new technologies connected to its edges and channel the viewer's sense of awe and curiosity back to the natural, fluid surroundings of the machine.

A recent competition sponsored by the National Grid in Great Britain received many creative forms to carry the cables—ninety-eight-foot-tall figures running with wires in their hands; deer carrying wires in their antlers; sails; insects; and a "flower tower." The winner of the contest was a sleek white "T-shaped" structure. One anticipates sublime power lines and other aesthetic structures. However, what may be proposed as an art form may also be rebranded as a marketing tool or to support a political ideology. One shudders to think that a real estate developer and politician like Donald Trump might tout the T-shaped tower design as part of the trillion-dollar infrastructure package that the nation needs and most Americans support. A massive, renewable electrification grid in any nation might have sweeping political impacts, but erecting thousands of giant *T*'s to carry power across the Midwest, often referred to in political terms as the "red states," would be an ironic realization of Stalin's declaration that "communism is Soviet Power plus electrification of the whole country." Neoliberal capitalism could mean investor-owned renewable infrastructure plus the government-mandated webbing of the whole country with high-voltage wires.

The National Grid contest also allows us to imagine, as Henry Dreyfuss did, varied creative designs for power lines. While some may correctly read corporate control or cronyism in the shape and design of the new towers, it is also possible that they could act as props for a cinemagraph: giant characters acting out a movement or scene as the driver or train passenger glides past. Or like Thoreau's telegraph harp, the lines might be positioned so as to produce music in certain seasons or times of the day. We might invite visitors to observe our metallic looms stretched toward the horizon or a valley with a colorful loom. The up-and-down flowing motion of the wires and the playful guys and pylons accentuate the stage.

Again, to support imaginative solutions to the power lines problem does not mean complacency with the grid we have inherited or acceptance of the corporate push to privatize, deregulate, and wield eminent domain for private gain. The success of such a poetics is not a given. On the one hand, to present facts or poems about these technically complex systems and some of the utilities' more egregious abuses may illicit a collective shrug—if the electricity is delivered, many would prefer not to think about where it begins or how it arrives or who owns and controls it. Bringing awareness to the lines may also backfire—those who had ignored the lines may begin to see them, realize the negative impacts, and work to bring down the grid. It may seem best to keep the lines muted, but our power lines, whether beauty or blight, cannot be ignored. They represent the difficult truths about electricity and landscape and the ways they shape our everyday lives.

If the first branch examines the lines as part of a movement through space, the second branch of a power lines poetics pauses to consider how these structures create place. Electric lines, like poetry, mediate exchanges of social practice. A published, printed page (paper or web based) often holds the lines of a poem in a static embrace. Electric lines rarely seem to be contained within a single, clearly defined place—they span place(s) and become part of a broader pattern. As architect David Leatherbarrow argues, "The task of landscape architecture and architecture, as *topographical arts*, is to provide the prosaic patterns of our lives with durable dimension and beautiful expression."[40]

If architects and artists can achieve this task, then to stand within the power-lined landscape would be to stand within what artist Tony Smith calls "something mapped out but not socially recognized." For Smith, who drove the unfinished New Jersey Turnpike in 1951, abandoned works and surrealist landscapes transmit something more than a function; they produce spaces outside tradition, an "artificial landscape without cultural precedent."[41] To offer a reading of one such unprecedented, artificial landscape, I would like to return to the place connected to the lines that inspired my childhood fascination, the lines positioned above and alongside Blondo Street.

The Final Level of Play

The Omaha Public Power District (OPPD) Arboretum on 108th and Blondo Street satisfies Aldo Leopold's goal of creating "a benchmark, a starting point, in the long and laborious job of building a permanent and mutually beneficial relationship between civilized men and civilized landscape" (fig. 25).[42] This arboretum's starting point was a massive substation designed and built in the 1970s on what had been a twenty-six-acre family farm. For most of the century this slice of Nebraska farmland produced corn, soybeans, wheat, and hay. The utility bought the entire parcel, but the substation's footprint only occupied a few of the twenty-six acres. The remaining land could not be resold to developers, and for decades the utility let everything around the substation and its transmission towers go to grass.

From outside the substation one can see mustard-yellow metal sheds, obelisk-shaped transformers, countless switches, lightning arrestors, and

25. Omaha Public Power District Arboretum, 2017. Photo by Daniel Wuebben.

circuit breakers with six sets of stacked discs splitting from bases the size of heavy metal dollies. The circuit breakers resemble giant metal cacti. All of this humming industrial stuff stands on a rock gravel pad surrounded by a barbed wire fence. One 220-kilovolt lattice steel tower is beside Blondo and carries power southwest toward Dodge Street; the other crosses diagonally across a portion of twenty-two acres around the substation.

In the 1990s Omaha Public Power District's lead arborist and a local landscape architect, John Royster, set about transforming the unused land into an arboretum. The goal was twofold: to offer a public green space and to teach visitors how to select and trim trees and shrubs so they would not interfere with power lines. For utilities across North America, keeping foliage away from their equipment is a costly and endless chore, especially in the summer months when cables expand, lines sag, and tree branches can cause arcs and flashes and trigger blackouts. In other words, for utilities trees often act like weeds—plants that grow where they are not wanted.

The OPPD Arboretum engages the public to be part of the utility's tree care solution. Among the 208 trees species and 201 shrub types, a series of educational displays teach visitors how to select and plant "the right tree in the right place." In addition, access to a gazebo, a small koi pond, rock walls, and wooden bridges has made the site popular with those seeking light exercise, walking their dog, or taking photographs. On warm weekend days dozens of photographers and their subjects spread across the arboretum and pose for baby photos, senior pictures, and family portraits. They enjoy this free, picturesque scenery, and based on photos uploaded to social media, they dutifully crop out the substation and overhead lines. The OPPD Arboretum indicates the potential power of digital images and social media to adjust, frame, and reevaluate how the public engages with the various layers of electric infrastructure. This landscaped and well-maintained green space represents the potential for a "permanent and mutually beneficial" relationship between the energy industry and environmental literacy.

The arboretum's most provocative space, and where I would begin to teach the layperson about infrastructure, may be the least technologically advanced (and least photographed). OPPD's substation plans did not include a patch of forest east of the old family farm. In the 1950s the Nebraska Department of Roads proposed an interstate bypass to wrap around Omaha and link back with the cross-country artery, Interstate 80. The engineers and planners considered building an off-ramp to Blondo Street. The plan never materialized, and other exits directly north and south satisfied the flow of traffic. For decades after the interstate was completed, this strip of land that could have been an off-ramp remained untouched.

I grew up in a house built in the 1970s on the opposite side of the interstate, a stone's throw from this forgotten pocket of wilderness. As Omaha spread westward, subdivisions, apartment complexes, shopping centers, and office parks transformed everything around this small forest and the OPPD substation. My mother warned us not to cross the interstate and go into the forgotten forest. When we did, we battled through tall grass and dense shrubs and found a few well-worn footpaths that led to remnants of campfires littered with beer cans, pornography, and anarchist graffiti— telltale markers of 1980s suburban delinquency. Decades later the Department of Roads donated this section of unused land to allow the arboretum to expand. Engineers and landscape architects paved wood chip walking paths through the forest, placed two wooden bridges across its shallow creek, and created an outdoor classroom with a few wooden benches. A local community college professor now regularly brings his botany students here for identification exams. I envision the outdoor classroom as an ideal place for conversations about the confluence of environmental history, transportation infrastructure, and electric power. Such a conversation could treat technology in its true mode, which Martin Heidegger says is "the mode of revealing." As technology is revealed, it aligns with the "bringing-forth" enacted in poetry and physics.[43]

In this old-growth forest it is difficult not to consider the ever-present crashing whorl of interstate traffic. That white noise was present in my home, and I often imagined the interstate was a seashore upon which thousands of metallic seashells were repeatedly beached and unzipped. These days I am not pulled into the grove by nostalgia for traffic sounds. Instead, I go to visit the eastern cottonwoods, Nebraska's state tree and one of the species *not* recommended for planting near power lines. In Willa Cather's Nebraska novel *O Pioneers!* a character muses that the Bohemians who first settled the land were "tree worshippers," to which another responds: "I like the trees because they seem more resigned to the way they have to live than other things do. I feel as if this tree knows everything I ever think of when I sit here." This eastern cottonwood, which once upon a time was surrounded by prairie grass or flat plains and is now bookended by an interstate and electric substation, still seems content to be. When I touch the rough, russet-colored bark of the three clustered cottonwoods that each rise eighty feet or more above earth's surface, I think of all the resigned bodies and systems beneath my feet. Some are decaying, churning into dust. Others are alive, burrowing for sustenance. The branches of mycelium and tree roots are the ancient precursors, perhaps the archetypes, for the wires in our walls and the electrical infrastructure overhead.

Underground roots have a similar shape and function as overhead transmission and distribution lines. Those bifurcating siphons tunnel into soft, loamy dirt, their longest fingers dug 40 or 50 feet below the earth's surface, their elbows ranging outward, some of them exposed on the banks of a nearby ditch. The other, metallic siphons are visible on the woodland's edge—aluminum and steel-reinforced cables hang between 45-foot tall wooden utility poles (which were once Douglas firs, a species harvested for Christmas trees). The network also shines on the clearing's western rise—quarter-inch lines locked to accordion insulators and attached to a 180-foot lattice steel tower.

This lattice steel tower and the exposed roots also serve as reminders: without these appendages the cottonwoods would perish; the grid would go dark. When they operate effectively, roots and lines are often forgotten. *Tree* and *electricity* attract warmer visions and more magnanimous associations. In this arboretum, and most landscapes across the country, attention is reserved for the blooming surprise: thick branches stem from this sturdy cottonwood trunk and reign over the bur oak, hackberry, and maple at their shoulders. Throughout the summer visitors will be drawn to this tree's thousands of waxy, silver-and-green leaves. As summer turns to fall, the leaves will turn bright-lemon, fade, and then drop to the forest floor. That image will grab attention. Visitors will not, unless prodded, turn to consider the field and the transmission tower. We tend to believe that nature is the snapshot of spectacular autumn foliage, not the gnarly ecosystem that feeds such colorful explosions.

Similarly, we are inclined to see electricity as a series of devices and icons: bolts with zigzagging lines, light bulbs aglow, rectangular wall sockets with vertical eyes and round nose that receive our plugs. Our motors, televisions, and various screens are electricity, not the wires hidden inside. Campaigns for energy-saving appliances and nimble electric vehicles with eco-conscious names such as Leaf, Bolt, Volt, and Tesla often ignore the massive, and sometimes environmentally damaging, energy networks these new technologies require to stay charged. The lines in the landscape are not often associated with the electric technologies in operation around us—and even upon and within us—every waking moment of our lives. Some of the electricity in that lattice steel tower, however, allows me to sit in my home across the interstate and type these lines.

To look at the eastern cottonwood and see an ecosystem is analogous to seeing that the electricity in those lines touches every aspect of our lives. To see the forest (as grid) for the trees (as transmission lines) is to appreciate infrastructure as a series of material artifacts, structures with a technological function *and* a social meaning. Their form, distinct from but entwined with their function, conveys a certain intention and agency. The lines, the trees, and all the values we might attach to them interconnect with landscape and create place. The lines transmit physical power; they also send powerful messages—messages that over the course of successive generations seem to shift, hide, evolve, and fluctuate.

Here, then, the romantic analogy linking underground roots and overhead power lines falls apart. The organic roots that sustain this forest spread radially from the tree trunk. Power lines rarely radiate from a center; they join a network that stretches so far and so deep that the grid they serve operates as its own kind of ecosystem. Unlike the swelling leaves above or the veiny roots that burrow below this rich Nebraska soil, the carefully engineered, owned, and managed metallic lines in the thawing field that drop toward the substation and parallel Blondo Street must hide in plain sight, like a bird on a wire. While they may not be natural, or entirely understood, those lines remain curious fixtures and icons that invite further untangling. The power line here is like a clear, thin glaze on the canvas of our historical, aesthetic, and technological milieu. Those webs that span the horizon and range beyond this landscape are electricity made visible; they occupy the razor-thin margin between the seen and unseen.

Acknowledgments

Some readers may be understandably exhausted by my use of electrical language and metaphors. They are advised to skip this section. A brilliant network of family, friends, and colleagues guided the composition of this text across many years and continents. I consider myself the humble engineer and line operator; my wife, Eva Hernandez, has been the executive behind the scenes. Eva, I am not sure if you have read a single word of this book during its winding path to print; however, probably more important, you have endured my countless detours, captured thousands of power line pictures for the cause and patiently shaped my view of life and language. Mi vida, mil gracias por tu apoyo.

My parents, Ted and Colleen, and my siblings, Cristen, Jenny, Jeremiah, Michaela, and Mary, have been lifelong sources of renewable energy. You each charged my childhood with love, laughter, and serious play. For me it is enough to imagine you reading this with mom in mind, patiently questioning my word choices and quietly beaming with pride.

Many friends have proven valuable conductors during my quest to find patterns in the sound. (In alphabetical order) Steve Alvarez, John Belitsky, Sarah Dodson, Chris Johnson, Mike Kelly, John Knuth, Michael Heim, Mark Malloy, Gerry McGill, Dana Naughton, Justin Oberst, Ryan Palmer, Lorcan Precious, and Jonathan Wegner each deserve to be recognized by name. It has been exciting to share my ideas at conferences, but possibly more rewarding has been the spontaneous, late-night talks and long road trips with these friends, during which I felt the spark of new ideas and saw them come to life through others' eyes. Special thanks to Adjua Gargi Nzinga Greaves, who within seconds of our first encounter challenged me to answer,

"What is beauty?" to which I responded, "Power lines." You repeatedly inspire me to pursue beauty, science, and poetry on this pale-blue dot. In the past few years, while composing this book, Colin Huerter, Keenan Krick, and Ian McElroy have been powerful insulators and spiritual advisors; our conversations have helped me contain stray voltage and to keep life humming along at a relatively healthy frequency. Brothers, my deepest gratitude.

This network's textual foundations were constructed during my studies at the City University of New York Graduate Center, where I relied upon the inexhaustible batteries of knowledge offered by Joan Richardson, David Reynolds, and the late Edmund Epstein. This book would not have been possible without Joan, who helped me gather the "clumsy jumps" into a steadier course of ideas. Teaching undergraduate composition and literature courses has made my writing and research possible, and the sparks to my pedagogical practice have been produced by Linda Adler-Kassner, Ammeiel Alcalay, Steve Alvarez, Doug Bradley, Stephen Casmier, Marlene Clark, Christopher Dean, Patricia Fancher, John Price, Madeleine Sorapure, and Olivia Walling.

Martin Woessner, of City College Center for Worker Education, and Todd Richardson and Eugene DiStefano, of the University of Nebraska Omaha (UNO), offered important advice on specific sections of this book, and I am lucky to call them colleagues and lifelong friends. My aunt and champion Kris Haller and the perspicacious Mark Malloy read the entire manuscript. They provided keen insights and encouragement that consistently amplified the message I wanted to send. All the mistakes and short circuits that remain are my own.

My experience working with Bridget Barry, Emily Wendell, and the University of Nebraska Press has been a delight, and I extend them my heartfelt thanks.

During the composition of the text, I enjoyed the support of an Urban Research Award from the University of Nebraska Omaha College of Public Affairs and Community Service as well as a UNO Research and Creative Activity mini-grant that helped defray the costs of revisiting the Southern California Edison Archive at the Huntington Library in San Marino, California.

Finally, Louise April and Micaela Willa, you are my sunshine and my power lines. I know I have missed some weekend adventures and bedtime talking stories in my scramble to finish this manuscript. You have made your mark on this book with thousands of welcome distractions and unscripted interruptions. In fact, here is Lou (age five), barging into my makeshift office in Grandma Juanita's apartment overlooking Calle Serafín in Segovia.

"Lou, I'm going to mention you in my book. What do you want me to say?"

"I don't know . . . wait! Good energy makes love!"

Amen.

Notes

Preface

1. James, *Principles of Psychology*, 273.
2. Morley and Robert. "Electric Fields."
3. Edwards, "The Spider Letter," *Jonathan Edwards Reader*, 1–8.
4. Emerson, *Essays and Poems*, 965.
5. Joyce, *Portrait of the Artist*, 17–18.
6. Nabokov, "A Matter of Chance," *Stories of Vladimir Nabokov*, 54.
7. "Capital Costs for Transmission Lines and Substations: Updated Recommendations for WECC Transmission Expansion Planning," Western Electricity Coordinating Council, prepared by Black & Veatch, February 2014, https://www.wecc.biz/Reliability/2014_TEPPC_Transmission_CapCost_Report_B+V.pdf.
8. Hall, *Out of Sight, Out of Mind Revisited*, 23–28.
9. "Power Outages Often Spur Questions around Burying Power Lines," *U.S. Energy Information Association*, July 25, 2012, http://www.eia.gov/todayinenergy/detail.php?id=7250.
10. On health issues, see "Electric and Magnetic Fields Associated with the Use of Electric Power," National Institute of Environmental Health Sciences, 2002; see also "Electric and Magnetic Fields," https://www.niehs.nih.gov/health/topics/agents/emf/index.cfm. For more on home values, see Bottemiller and Wolverton, "Price Effects," 45–62. On the dangers to habitat and wildlife, see Loss, Will, and Marra, "Refining Estimates of Bird Collision and Electrocution Mortality," e101565.
11. Eric Tucker and Chris Kahn, "Easy Fix Eludes Power Outages Problem in the United States," *Associated Press Archive*, July 4, 2012.
12. Thoreau, *Journal*, 4:458–59.

13. "The Magnetic Telegraph," *Guardian of Health* 3, no. 3 (1848), Nineteenth Century Collections Online, tinyurl.galegroup.com/tinyurl/85v5E0. The anecdote also appears in the chapter "Strange Notions Concerning the Telegraph," in *The London Anecdotes for All Readers*, ed. Charles Archer (London: Bogue, 1848).

Introduction

1. Morse, *Letters and Journals*, 2:6.
2. Morse, *Letters and Journals*, 2:85.
3. Morse, *Letters and Journals*, 2:210.
4. Huurdeman, *Worldwide History of Telecommunications*, 61.
5. U.S. Bureau of the Census, *Telephones and Telegraphs: 1902*, 99.
6. The data about telegraph wire mileage and placement can be found in "Use 15,000,000 Miles of Message Wires," *New York Times*, August 9, 1909. Also note the observation from the U.S. Bureau of Census in 1902: "In 1902 the telegraph business was practically controlled by two companies yet in spite of the tendency of consolidation to reduce the number of lines and offices, the mileage of wire in operation was more than four times and the number of messages nearly three times greater than in 1880. The wire mileage in operation in 1902 exclusive of 16,677 nautical miles of cable was 1,027,137 miles greater than in 1880." U.S. Bureau of the Census, *Telephones and Telegraphs: 1902*, 99.
7. Phillip Hone's review of Morse's work, qtd. in Larkin, *Samuel F. B. Morse*, 114.
8. "Professor Morse's Great Historical Picture," *Yankee Doodle*, October 10, 1846, 5.
9. For more on how Morse's aesthetic training entwined with his technological innovations, see Gillespie, *Early American Daguerreotype*.
10. The analogy of the "chain" between environmental history and history of technology appears in Russell et al., "Nature of Power," 248.
11. *New York Herald*, May 30, 1844.
12. Czitrom, *Media and the American Mind*, 13.
13. Major studies of the history of electric technologies and electrification on American culture include French, *When They Hid the Fire*; Hochfelder, *Telegraph in America*; Hughes, *Networks of Power*; Jonnes, *Empires of Light*; Nye, *Electrifying America* and *American Technological Sublime*; and Simon, *Dark Light*.
14. Otis, *Networking*; Delbourgo, *Most Amazing Scene of Wonders*; Halliday, *Science and Technology in the Age of Hawthorne*, 3; Menke, *Telegraphic Realism*; Gilmore, *Aesthetic Materialism*; Bazerman, *Languages of Edison's Light*; Lieberman, *Power Lines*, 5.

15. Glaser, "Nice Towers," 23–24.
16. Levy, "Aesthetics of Power," 575–607.
17. Mitchell, *Landscape and Power*, 5, 2.
18. Definition of *wilderness* made by the National Wilderness Preservation System. Qtd. in Nash, *Wilderness and the American Mind*, 5.
19. Marx, *Machine in the Garden*, 128. For more on the role of nature writing and landscape in Jefferson's politics, see Miller, *Jefferson and Nature*.
20. Leopold, "Conservation Esthetic," 109.
21. Conron, *American Landscape*, xviii.
22. Olson, *Collected Prose*, 17.
23. Hughes, *Networks of Power*, 1.
24. Neil Armstrong, *Century of Innovation*, vii.
25. Hughes, *Networks of Power*, 1, emphasis mine.
26. Hughes, "Evolution of Large Technical Systems," 52.
27. Hughes, "Technological Momentum," 108.
28. Mumford, *Technics and Civilization*, 223.
29. Bakke, *Grid*, xxvi.

1. Wires in the Garden

1. Estimates of total miles of telegraph wire for 1853 and 1866 appear in U.S. Bureau of the Census, *Historical Statistics of the United States* (1960), chap. R, 475, 484.
2. Marx, *Machine in the Garden*, 203.
3. Marx, *Machine in the Garden*, 28.
4. Whitman, *Complete Poetry and Collected Prose*, 531; Marx, *Machine in the Garden*, 223.
5. Mumford, *Technics and Civilization*, 12.
6. Thoreau, *Week on the Concord and Merrimack Rivers*, 177.
7. Benjamin French, "The Changes of the World," *New-Hampshire Statesman and State Journal*, February 7, 1845.
8. The phrase *Morse lines* often referred to lines built based on Morse's patented devices—as opposed to the *House lines* (for Royal E. House) and *Bain lines* (for Alexander Bain). Morse's system became so pervasive that *Morse lines* became synonymous with all telegraph lines. See, e.g., Shaffner, *Telegraph Companion*.
9. Bryant, *Prose Writings*, 281.
10. Morse, *Lectures on the Affinity of Painting*, 24. Nicolai Cikovsky Jr. suggests much of Morse's writing on landscape was unoriginal: "[Morse's] consultation of encyclopedias and other reference books and his close, extensive, and

not always clearly announced paraphrases, particularly of Blair's *Rhetoric* and Whatley's *Modern Gardening*, may still be other signs of his attempt to give the impression of wider learning than he actually possessed."

11. Morse, *Lectures on the Affinity of Painting*, 50.

12. Bryant, *American Landscape*.

13. Morse, "Academies of the Arts," 23.

14. Cole, "Essay on American Scenery," 4, 11.

15. Emerson, *Essays and Poems*, 96.

16. A more thorough account of Emerson's readings in electrical science and use of electric language can be found in Wilson, *Emerson's Sublime Science*.

17. Emerson, "Nature," *Essays and Poems*, 10.

18. Emerson, "The Poet," *Essays and Poems*, 467.

19. Whitman, *Complete Poetry and Collected Prose*, 55.

20. Whitman, *Complete Poetry and Collected Prose*, 56, 387.

21. Morse, *Letters and Journals*, 2:31.

22. "In the Midst of a Revolution," *New York Herald*, 158, June 6, 1844.

23. O'Sullivan, "Annexation," 9. O'Sullivan's connection between manifest destiny and a "nervous system of magnetic telegraphs" is shown in Hietala, *Manifest Design*, 197.

24. Qtd. in Czitrom, *Media and American Mind*, 12.

25. *New York Evangelist* 17, no. 41, October 8, 1846.

26. For more on journalism and the telegraph, see Hochfelder, *Telegraph in America*; and Blondheim, *News over the Wires*.

27. Alexander Jones, *Historical Sketch*, 77–81.

28. Huurdeman, *Worldwide History of Telecommunications*, 65.

29. "The Electric Telegraph in the Scioto Valley," *Scioto Gazette* 27, October 4, 1848.

30. Blondheim, "Three Phases," 80.

31. "The End of Fanaticism in Boston," *New York Herald*, 347, December 17, 1845.

32. *Charleston Daily Courier*, June 8, 1846; *New Orleans Commercial Times*, July 13, 1846. Both citations qtd. in Blondheim, "Three Phases," 80–81.

33. *Mississippi Creole*, February 16, 1850.

34. Turnbull, *Electro Magnetic Telegraph*, 256.

35. Huurdeman, *Worldwide History of Telecommunications*, 63.

36. Howe, *What Hath God Wrought*, 695.

37. U.S. Bureau of the Census, *Historical Statistics of the United States: Colonial Times to 1970*, chap. R: Communications.

38. Holmes, *Writings of Oliver Wendell Holmes*, 8:7.

39. Ayers, *What Caused the Civil War*, 141.
40. For more on the National Telegraph Act, see John, *Network Nation*, 116–25.
41. Thompson, *Wiring a Continent*, 440.
42. Schivelbusch, *Railway Journey*, 29.
43. "Curiosities of Railway Travel," *Harper's Magazine* 4, no. 30, April 12, 1851.
44. Schivelbusch, *Railway Journey*, 31.
45. Czitrom, *Media and the American Mind*, 11.
46. Hochfelder, *Telegraph in America*, 74.
47. "Editor's Table: Exhibition of the National Academy of Design," *Knickerbocker New York Monthly Magazine* 42, no. 1 (July 1853): 95.
48. "Dr. Alexander Jones: Obituary," *New York Herald*, August 26, 1863.
49. Jones, *Historical Sketch*, v–vi.
50. Benjamin French, "Changes of the World" in, *The Telegraph Dictionary and Seaman's Signal Book* (Baltimre: F. Lucas Jr., 1845), v.
51. Turnbull, *Electro Magnetic Telegraph*, v.
52. Turnbull, *Electro Magnetic Telegraph*, 22
53. Review of Turnbull, *North American and United States Gazette*, November 26, 1853.
54. Hawthorne, *House of Seven Gables*, 253.
55. Hawthorne, *House of Seven Gables*, 96, 264, 254.
56. Qtd. in Czitrom, *Media and American Mind*, 12.
57. Hawthorne, *House of Seven Gables*, 264.
58. Hawthorne, *House of Seven Gables*, 264.
59. Emerson, "The Poet," *Essays and Poems*, 455.
60. Hawthorne, *House of Seven Gables*, 264.
61. Hawthorne, *House of Seven Gables*, 254.
62. Melville, *Moby-Dick*, 394, 310.
63. Melville, *Moby-Dick*, 281.
64. Thoreau, *Week on the Concord*, 394.
65. Thoreau, *Week on the Concord*, 394, emphasis mine.
66. Thoreau, *Journal*, 2:344. Semaphore telegraphs transmitted messages through a series of towers resembling windmills. The wooden blades or shutters could be arranged to connote a certain letter or word, which was seen and copied by the signalman at the next tower and so on until the message reached its final destination. In the 1790s the French central government began operating a semaphore telegraph between Paris, Strasbourg, and Lille. In the 1830s the U.S. government considered building an optical telegraph along the East Coast, but the plans never materialized, and Morse's invention soon made the semaphore telegraph obsolete.

67. Thoreau, *Journal*, 2:338.

68. Thoreau, *Week on the Concord*, 177.

69. Thoreau, *Journal*, 3:442.

70. Thoreau, *Journal*, 2:450.

71. S. M. Partridge, "The Magnetic Age," *Knickerbocker New York Monthly Magazine* 28, no. 6 (December 1846).

72. "An Evening with the Telegraph Wires," *Atlantic Monthly* 2, no. 11 (September 1858): 489–95, 491, 495.

73. Qtd. in Czitrom, *Media and American Mind*, 8.

74. *Magazine of Science* (London), pt. 53 (October 1847): 140.

75. Thoreau, *Journal*, 2:496–97.

76. Thoreau, *Journal*, 3:11, 71.

77. Thoreau, *Journal*, 4:342.

78. Thoreau, *Week on the Concord*, 568, 566.

79. Gamble, "Early Reminiscences of the Telegraph on the Pacific Coast," 321–26.

80. Gamble, "Wiring a Continent."

81. Brown, *1861 Diary*, 90–91.

82. Jepsen, "Telegraph Comes to Colorado," 1–23.

83. Ware, *Indian War of 1864*, 80–81.

84. *Daily Alta California* 17, June 21, 1865.

2. Frontier Lines and Telegraph Forests

1. Nye, *American Technological Sublime*, 152.

2. Qtd. in Irwin, *New Niagara*, 110.

3. "20,000 Horsepower over One Wire," *New York Times*, May 24, 1893.

4. Qtd. in Adams, *Niagara Power*, 2:164.

5. Stringer, *Wire Tappers*, 112.

6. Stringer, *Phantom Wires*, 26.

7. Nye, *Electrifying America*, 8.

8. Schatzberg, "Culture and Technology in the City," 69.

9. Bogue, *Frederick Jackson Turner*, 91; Faragher, "Introduction," 1.

10. Pomeroy "Toward a Reorientation of Western History," 579–600.

11. Pursell, *Machine in America*; Cronon, *Nature's Metropolis*, 34.

12. Turner, "Significance of the Frontier," 39.

13. Turner, "Significance of the Frontier," 4.

14. Slotkin, *Gunfighter Nation*, 55–87.

15. Slotkin, *Gunfighter Nation*, 37.

16. Faragher, "Introduction," 2.

17. "Electricity at the Fair," *Harper's Weekly*, July 16 1892, 693.

18. Adams, "Promotion of New Technology through Fun and Spectacle," 50.

19. For more on the Electricity Building at the Columbian Exposition of 1893, see also Johnson, *History of the World's Columbian Exposition*, 297–99; Murat Halsted, "Electricity and the Fair," *Cosmopolitan*, May 1893, 577–82; and Barrett, *Electricity at the Columbian Exposition*, 16–19. Another excellent resource about the Chicago World's Fair has been created by the UCLA History Department and the urban simulation team.

20. William James did not attend the fair but wrote to his brother Henry: "*Everyone* says one ought to sell all one has and mortgage one's soul to go there, it is esteemed such a revelation of beauty . . . People cast away all sin and baseness, burst into tears and grow religious under the influence!" William James to Henry James, September 22, 1893, Skrupskelis and Berkeley, *Correspondence of William James*, 280.

21. With over 21.5 million paid admissions, somewhere between 5 and 10 percent of the population of the United States visited the fair. For more on attendance, see Badger, *Great American Fair*, 109.

22. "History Making Celebration of the Only Electrical Banquet the World Has Ever Seen," *Buffalo Evening News*, January 13, 1897, 1:1–2; 4:2–5.

23. Of Nikola Tesla one journalist wrote: "The men living at this time who are more important to the human race than this young gentleman can be counted on the fingers of one hand; perhaps on the thumb of one hand." *New York Sun*, March 14, 1895, 6.

24. *New York Times*, July 16, 1895, 10:5.

25. "History Making Celebration," 1:1–2; 4:2–5.

26. Tesla, "On Electricity." I quote from the piece as published in *Nikola Tesla: Lectures, Patents, Articles*, 101–8. A transcript of the speech appeared in the *Electrical Review*, January 27, 1897.

27. Tesla, "On Electricity," 103.

28. Tesla, "On Electricity," 106, emphasis mine.

29. Tesla, "On Electricity," 106.

30. Adams, *Niagara Power*, 271, 357, 362, 353, 355.

31. Tesla, "On Electricity," 107.

32. Francis Lynde Stetson, "The Use of the Niagara Water Power," *Cassier's Magazine* 8 (1895): 192.

33. "Power at Niagara Falls," *New York Times*, February 11, 1900.

34. Cole, "Essay on American Scenery," 12.
35. *Niagara Falls Gazette*, July 15, 1885, qtd. in Irwin, *New Niagara*, 79.
36. Nye, *American Technological Sublime*, 135.
37. Irwin, *New Niagara*, 110.
38. "The buildings and grounds of the Exposition are lighted by electricity. About 8,000 arc lamps of 2,000 candle power and about 130,000 incandescent lamps of sixteen candle power are required. Besides this from 3,000 to 3,500 horse power is required for the operation of the machinery of the exhibitors. To furnish and transmit this 24,000 horse power the Exposition Company has constructed a plant which, though a complete station itself, is composed of a number of smaller complete plants." White and Ingleheart, *World's Columbian Exposition*, 476.
39. Adams, *Niagara Power*, 370.
40. "Electric Power at Niagara," *New York Times*, January 11, 1903. Special to the *New York Times*, originally published in Washington, January 10, 1903. In addition, *The World Almanac and Encyclopedia 1901* reported: "The Niagara Falls Power Company made a second large extension of its plant at Niagara Falls, New York, during 1901 and is now the largest electric transmission company in the world. The General Electric Company whose extensive plant is at Schenectady, New York, receives its power from a water power plant located at Mechanicville some miles away. The numerous electric transmission plants throughout the West are proving of tremendous benefit to the surrounding territory and have produced profitable returns for their stockholders" (189).
41. "Niagara Put in Harness," *New York Times*, July 7, 1895.
42. "Mr. Hill Gets the Prize," *New York Times*, February 7, 1891.
43. Adams, *Niagara Power*, 203, 269.
44. Billington and Billington, *Power, Speed, and Form*, 27–30.
45. Irwin, *New Niagara*, 107.
46. For more on Michael Dolivo-Dobrowolsky's AC system and Nikola Tesla's conflicts about the creator of the AC system, see Seifer, *Wizard*, 73–74, 80–81. Hughes, *Networks of Power*, 133.
47. In spring 1893 the Westinghouse Company accused some of its employees of stealing blueprints and data relating to its Niagara proposal and selling them to representatives of General Electric. For more on the accusations and legal battle involving the theft of these documents, see "Electric Companies at War," *New York Times*, May 7, 1893.
48. Hughes, *Networks of Power*, 139; Adams, *Niagara Power*, 271–72.
49. Adams, *Niagara Power*, 182.

50. "The Niagara Falls Electric Line," *New York Times*, August 8, 1894.

51. "Distribution of Niagara's Power," *New York Times*, May 6, 1894.

52. Stetson, "Use of the Niagara Water Power," 192.

53. Edison and General Electric, the obvious losers in the "Battle of the Currents," were given the contract to build transmission lines capable of transmitting 11,000 volts from Niagara to Buffalo. The lines built by General Electric carried an unprecedented amount of the Tesla- and Westinghouse-generated two-phase alternating current. When the electricity reached its destination, it was transformed into a lower voltage and used to power motors, electric railcars, and streetlights. *Review of Reviews*, August 12, 1895, 208–9.

54. H. G. Wells, "End of Niagara," *Harper's Weekly* 50, July 21, 1906, 1019.

55. Adams, *Niagara Power*, 277, 278.

56. *Electrical Engineer*, May 27, 1896, 577–79.

57. "The National Electrical Exposition at the Grand Central Palace, New York," *Scientific American*, May 16, 1896, 309–10.

58. "Street Dangers: The Rotten Telegraph-Pole: Conclusion of the Inquest in the Case of Ann M'Guire—Censures and Recommendation of the Jury," *New York Times*, August 12, 1876.

59. "Rotten Telegraph Poles and Trees," *New York Times*, August 18, 1876.

60. "The Telegraph Forest," *New York Times*, August 12, 1876.

61. Joseph J. Cunningham, *New York Power*, 9, 22.

62. John Trowbridge, "Progress of Electrical Science during 1883," *Science* 29 (February 1884): 258–61.

63. *Electrical World*, November 2, 1889, 292–93.

64. Sullivan, "Fearing Electricity," 9.

65. "The Pole and Wire Nuisance," *New York Times*, June 26, 1884.

66. "Telegraph Poles Must Go," *New York Times*, September 19, 1884.

67. "The Underground Wire Bill," *New York Times*, June 14, 1885.

68. "Wires Illegally Strung," *New York Times*, September 18, 1886.

69. New York Board of Electrical Control, qtd. in Nye, *Electrifying America*, 48.

70. Casson, *History of the Telephone*, 127–28.

71. "To Avoid Lost Land Lines," *New York Times*, March 8, 1888.

72. "Bury the Wires," *New York Times*, May 12, 1888.

73. Tesla, "New System of Alternate Current Systems and Motors," 10.

74. Essig, *Edison and the Electric Chair*, 146. Essig also provides a thorough discussion of Edison, the battle of the currents, and the use of animals for electrical experiments (134–62).

75. Sullivan, "Fearing Electricity," 8–16.
76. "City and Suburban News," *New York Times*, February 4, 1892.
77. Weitenkampf, *Manhattan Kaleidoscope*, 52.
78. Cable, *Blizzard of '88*, 186.
79. "Public Works Department; Mr. Gilroy's Report of Its Operations Last Year," *New York Times*, January 1, 1893.
80. Schuyler Wheeler, "Electric Lighting in New York," *Harper's Weekly*, July 20, 1889, 593–96.
81. "The Trolley Street-Car System" *Harper's Weekly*, August 6, 1892, 747.
82. Bakke, *Grid*, 50.
83. Robinson, *Better Binghamton*, 45; Robinson, *City Plan for Raleigh*, 34.
84. A chapter devoted to Thayer's *Wired Love* can be found in Otis, *Networking*, 147–79. Mark Twain's *Connecticut Yankee* is most thoroughly analyzed in Lieberman, *Power Lines*. Henry James's *In the Cage* (1898) is discussed in Richard Menke's article "Telegraphic Realism" and subsequent book, *Telegraphic Realism*; Gilmore's *Aesthetic Materialism*; and Sam Halliday's *Science and Technology in the Age of Hawthorne*. Discussions of the telegraph in Frank Norris's novel *The Octopus* (1901) include Halliday, *Science and Technology in the Age of Hawthorne*, 20–23; MacDougall, "Wire Devils," 74; and Berte, "Mapping 'The Octopus,'" 202–24.
85. Stringer, *Wire Tappers*, 248.
86. Stringer, *Wire Tappers*, 39–40.
87. Stringer, *Wire Tappers*, 5, 6, 149, 220.
88. Stringer, *Wire Tappers*, 39, 40, 39–40, 115.
89. "Scientific Wire Tappers Who Band Together and Work the Pool Rooms," *National Police Gazette* 80, no. 1284, March 29, 1902.
90. Stringer, *Wire Tappers*, 237.
91. Yates, "Telegraph's Effect on Nineteenth Century Markets and Firms," 149–63.
92. "Cotton Men in a Panic: New Orleans Exchange Suspends on Receiving False Quotations. Wire Tapping Is Charged," *New York Times*, September 30, 1899.
93. Other treatments of market manipulation include the Frank Norris's Progressive Era novel *The Pit* (1903) and the silent film it inspired, D. W. Griffith's *Corner in Wheat* (1909).
94. The corruptive influence of the cotton exchange and telegraph reports of price fluctuations are also central components in William Faulkner's *Sound and the Fury* (1929).
95. Stringer, *Phantom Wires*, 26.
96. Stringer, *Phantom Wires*, 228.

97. Stringer, *Phantom Wires*, 153.

98. The data about telegraph wire mileage and placement can be found in "Use 15,000,000 Miles of Message Wires," *New York Times*, August 9, 1909. Also note the observation made in the U.S. Bureau of the Census, *Telephones and Telegraphs: 1902*: "In 1902 the telegraph business was practically controlled by two companies yet in spite of the tendency of consolidation to reduce the number of lines and offices, the mileage of wire in operation was more than four times and the number of messages nearly three times greater than in 1880. The wire mileage in operation in 1902 exclusive of 16,677 nautical miles of cable was 1,027,137 miles greater than in 1880" (99).

3. Wood Poles, Steel Towers, Modernist Pylons

1. Worster, *Under Western Skies*, 87.

2. Vincent, "Interconnected Transmission System of California," 565; Hughes, *Networks of Power*, 271.

3. The decision to set the Mill Creek generators to 50 cycles per second (50Hz instead of 60Hz) would make this part of Southern California an "electrical enclave" until the 1940s. For more on the impact of this decision, see Nathan Masters, "Before 1948, LA's Power Grid Was Incompatible with the Rest of the U.S.," *Gizmodo.com*, February 4, 2015.

4. Nye, *Electrifying America*, 32.

5. *Thirteenth Census of the United States, 1910: Population*, https://archive.org /details/thirteenthcensus00hunt.

6. "Southern California: The Story of Its Development, Resources, and Progress," *Harper's Weekly*, April 2, 1904, 505–6.

7. J. E. MacDonald, "Joint Pole Lines in Los Angeles," *Electrical World* 53, no. 17, 1909, 965–67.

8. MacDonald, "Joint Pole Lines," 965–67.

9. Bowser, *Biograph Bulletins, 1908–1912*, 284.

10. *The Lonedale Operator* was filmed January 14–16 and February 2–4, 1911. It was shot in Los Angeles Studio and Inglewood, California. For more on the filming schedule, see Kristin Thompson's entry on "The Lonedale Operator," 18.

11. Knight, *Liveliest Art*, 32.

12. Griffith, *Man Who Invented Hollywood*.

13. For full shot-by-shot analysis of *Lonedale Operator*, see Thompson, "The Lonedale Operator"; Bellour, "To Alternate/Narrate"; and Gunning, "Systematizing the Electric Message."

14. Young, "Media on Display," 229, 231.

15. Gunning, "Systematizing the Electric Message," 27.

16. Bellour, "To Alternate/Narrate," 262, 273.

17. The visual spectacle of these two technologies is also at the heart of two of the most famous moving images of early cinema, Louis Lumière's *Arrival of the Train* (1895) and the pistol-pointing cowboy close-up of Edwin S. Porter's *Great Train Robbery* (1903). Both of these famous clips show technology making a sudden entrance and turning toward the camera and thus the audience.

18. Porter's *Great Train Robbery*, which some scholars say is the first western film, initiated the trope of the isolated, defenseless telegraph operator. In the opening shots of that film, two robbers enter a telegraph office, beat the station operator, and tie him to his chair. Lusted, *Western*, 74.

19. Halliwell, *Halliwell's Who's Who*, 176.

20. "The Transmission Systems of the Great West," *Electrical World* 59, June 1, 1912, 1143.

21. *Sacramento Union*, May 18, 1914.

22. In 1879 the California Electric Company (now PG&E) used two direct current generators from Charles Brush's company to supply multiple customers in San Francisco with power for their arc lamps. This was the first case of a utility selling electricity from a central plant to multiple customers via distribution lines.

23. John Wesley Powell, "The Non-Irrigable Lands of the Arid Region," *Century Magazine* 39 (April 1890): 915–22.

24. Williams, *Energy and the Making of Modern California*, 176.

25. "Power Sent by Wire: California Pioneered This Field," *San Francisco Chronicle*, July 15, 1895.

26. "Electric Power to Turn the Wheels of Progress," *Weekly Chronicle* (San Francisco), July 25, 1895.

27. *San Francisco Examiner*, June 3, 1895, 27.

28. "Prospects of Electrical Engineering on the Pacific Coast," *San Francisco Call*, June 1, 1895, reprinted in *Electrical West* 1 (July 1895): 27–28.

29. Williams, *Energy and the Making of Modern California*, 178.

30. "Electric Power Transmissions of Southern California," 45–47.

31. Williams, *Energy and the Making of Modern California*, 187.

32. A. H. Babcock, "An Electrical Prophecy," *San Francisco Chronicle*, December 30, 1894, 17. Other futuristic visions of the age, such as Bellamy's *Looking Backward*, described "house-wire" systems that would provide music on demand, but the "Electrical Prophecy" is unique in its mention of the poles in the landscape that would be required to deliver electricity.

33. Low, "World's Longest Electric Power Transmission," quotation on 160–61.

34. Drew, "1901 Transmission Line," 80–88.

35. "Longest Electrical Power Transmission," *New York Times*, January 19, 1902.

36. Low, "World's Longest Electric Power Transmission," 145.

37. French, *When They Hid the Fire*, 5.

38. Hughes, *Networks of Power*, 266.

39. "Day of Electricity Comes," *Sacramento Union*, no. 52, December 22, 1915.

40. T. C. Martin, "The Longest Power-Transmission Line in the World," *American Monthly Review of Reviews* (March 1902): 305.

41. Hunter and Bryant, *History of Industrial Power*, 3:358.

42. T. C. Martin, "The Harnessed Hudson," *American Monthly Review of Reviews* (1903): 200.

43. *Sacramento Union*, no. 37, June 6, 1920.

44. James C. Williams, "Otherwise a Mere Clod: California Rural Electrification," *IEEE Technology and Society Magazine* (December 1988): 14, 15.

45. In addition to the lines stemming from Santa Ana and Borel, electricity arrived into Los Angeles on the 118-mile, 75Vv line from Kern River No. 1 in 1907; the 239-mile, 95kV line from Bishop Creek in 1911; and the 241-mile, 140kV line from Big Creek in 1913. The Big Creek system would provide the majority of Southern California's power for almost three decades.

46. *Los Angeles Sunday Times*, November 9, 1913.

47. Brigham, *Empowering the West*, 127.

48. "Kern River Transmission Line Construction: Power Plant of the Edison Electric Company, Los Angeles, California," *Electrical World* 50, no. 9, August 31, 1907, 401–2.

49. Jones, *Routes of Power*.

50. "Kern River Transmission Line Construction," 401–2. Detailed descriptions of the other features of the Kern River system appeared in 1907 in previous issues of *Electrical World* on August 10 (50, no. 6, 277–81), August 17 (no. 7, 317–22), and August 24 (50, no. 8, 359–63).

51. "The Kern-1 Line 1907: M-4800 'Largest Insulator Used for Long-Distance Transmission.'" *R-infinity.com* (Winter 2002–3), http://www.r-infinity.com /Kern_M-4800/index.htm.

52. Benson, "Generating Infrastructural Invisibility," 103–30.

53. Bakke, *Grid*, 117–50. Bakke describes how both pesky squirrels and overgrown trees can short out power lines in an excellent chapter entitled "Things Fall Apart."

54. Benson, "Generating Infrastructural Invisibility," 120.

55. *Southern California Edison Quarterly Newsletter*, "Preparing for Six Million Population," *Southern California Edison Archive*, Huntington Library, box 381, folder 8.

56. "The Adonis Pioneer," Southern California Edison Archive, box 383, folder 1, Myers, Publicity and Publications File, Huntington Library, San Marino.

57. "Profiles: Artist in a Factory," *New Yorker*, August 29, 1931.

58. Robert Coe and D. St. Laurent, "Dreyfuss Towers Carry 220-kV along Boulevard," *Electrical World*, January 6, 1969, 23–24.

59. "Towers Win Top Prize in National Contest," *Desert Sun*, no. 214, April 11, 1969.

60. Roger Roark, "12 Jolly Green Giants March to Disney World," *Electrical World* (December 1971): 54–55.

61. Goulty, *Visual Amenity Aspects of High Voltage Transmission*, 180, lists the various Dreyfuss designs built in the 1970s: 138kV steel pole line for Toledo Edison Co. USA in July 1973; 765 kV steel H frame for Niagara Mohawk USA Transmission Distribution, June 1973, 67; 230kV steel pole for Philadelphia Electric Company Transmission and Distribution, June 1973, 83; 500kV Cor-Ten steel pole for Tennessee Valley Authority, Transmission and Distribution, February 1976, 53; 230kV steel pole for Ontario Hydro Canada Span, June 1975, 11; and seven Y-shaped 500kV towers on the last one and half mile–section of the Public Service Electric and Gas Co. line between Branchburg and Whitpain in the United States (108).

62. *Electrical World and Engineer* 39, no. 5, February 1, 1902, 188.

63. Bennett, "Aesthetic Considerations Affecting Power Development," 117, 119.

64. Bennett, "Aesthetic Considerations Affecting Power Development," 118.

65. Both poems by Auden and Spender and further analyzed in James Purdon's article "Electric Cinema, Pylon Poetry." Purdon effectively argues that the pylons that appeared in the British landscape in the late 1920s provided a "site of rhetorical struggle in which components could be coded as revolutionary or reformist, socialist or imperial, utopian or corrupting." *Amodern 2: Network Archaeology*, 2013, http://amodern.net/article/electric-cinema-pylon-poetry/.

66. Flinchum, *Henry Dreyfuss*, 174.

67. For reviews of the film *America*, see Schickel, *D. W. Griffith*, 492–93.

68. Flinchum, *Henry Dreyfuss*, 29, 32.

69. Alva Johnston, "Nothing Looks Right to Dreyfuss," *Saturday Evening Post* 22 (November 1947): 131–39.

70. U.S. Bureau of the Census, *Historical Statistics of the United States, Colonial Times to 1970* (Washington DC, 1976), 828.

71. Interview with Southern California Edison Archive vice president W. E. Montgomery in "What Is a Feasible Program for Converting to Underground?" *Electrical World* 18 (September 1967).

72. Gordon D. Friedlander, "Esthetics and Electric Energy," *IEEE Spectrum* (1965).

73. *New York Times*, June 8, 1965, 40.

74. Levy, "Aesthetics of Power," 575–607.

75. "Utilities Seek Prettier Power Lines," *New York Times*, July 29, 1966.

76. "Beautification Fever High Electric Utilities' Towers," *Wall Street Journal*, July 28, 1988.

77. "From Pole to Pole," *Architectural Forum* (April 1968): 101.

78. "A Designer Talks about Transmission Towers," *Electrical World*, February 20, 1967, 67–69.

79. "Designer Talks about Transmission Towers," 67–69.

80. Dreyfuss, *Electric Transmission Structures*, 9.

81. Dreyfuss, *Electric Transmission Structures*, 9.

82. Jordan Lummis, "Progress Report of Esthetic Designs for Transmission Structures Research Project," Minutes of the Transmission and Distribution Committee, Edison Electric Institute, Washington DC (January 1968), qtd. in Levy, "Aesthetics of Power," 600.

83. Susan Tikalsky and Cassandra Willyard, "Aesthetics and Public Perceptions of Transmission Structures," *Right of Way EPRI* (2007): 34–38.

84. "There's No Way to Beautify a Power-Line Pylon," *New York Times*, September 9, 1982.

85. For more on myths surrounding Tesla and "free energy," see Carlson, *Tesla*, 347.

4. Public Perceptions and Power Line Battles

1. Cohn, *Grid*, 80.

2. Nye, *Consuming Power*, 183.

3. Wasik, *Merchant of Power*, 232.

4. "Public Utility Holding Company Act of 1935: 1935–1992," Energy Information Administration, U.S. Department of Energy (January 1993): 6, https://www.eia.gov/electricity/archive/0563.pdf.

5. Interview with Orson Welles, qtd. in Naremore, *Orson Welles's Citizen Kane*, 21.

6. Norris, *Fighting Liberal*, 248, 318.

7. U.S. Bureau of the Census, *Agriculture: 1950*.

8. Rural Electric Cooperative Association, *Next Greatest Thing*, 121.

9. Rural Electric Cooperative Association, *Next Greatest Thing*, 2, 87.

10. "'The Spanish Earth' Is a Plea for Democracy," *New York Times*, August 21, 1937.

11. Whitman, *Complete Poetry and Collected Prose*, 531.

12. *Hydro: Power to Make the American Dream Come True*, Bonneville Power Administration, 1939, https://www.youtube.com/watch?v=Oafnr0J4a5I.

13. Guthrie, *From California to New York Island*, 31.

14. Glaser, *Electrifying the Rural American West*, 9.

15. Kline, *Consumers in the Country*, 161.

16. Qtd. in Kline, *Consumers in the Country*, 162; original in John Carmody to Benton Rural Electric Association, WA-83, August 12, 1938.

17. *Historical Statistics for the United States, Colonial Times to 1970*, chap. S, "Energy," 820.

18. 1970 National Power Survey of the Federal Power Commission, I-13-4.

19. Furby et al., "Public Perceptions of Electric Power Transmission Lines," 22.

20. Karady, "Environmental Impact of Transmission Lines"; and Gretchen McKay, "Taliesin West," *Pittsburgh Post-Gazette*, March 27, 2010.

21. "New Con ED Line Designed to Blend with Land Upstate," *New York Times*, January 2, 1969, 33.

22. "New Route for 500-kV Line Eases R-O-W Tiff," *Electrical World*, July 1, 1963, 61.

23. "Potomac Edison Losing Battle of Antietam," *Electrical World*, August 7, 1967, 53–54.

24. Decades later the ability to buy and sell kWh in an open, real-time market allowed Enron to manipulate electricity prices, which contributed to the western U.S. energy crisis of 2000 and 2001. Furthermore, the same automatic relays that redistribute thousands of megawatts of power in the case of an outage contributed to the cascading failure of power plants and the Northeast blackout of 2003.

25. 1970 National Power Survey, I-13-6.

26. Carson, *Silent Spring*, 25.

27. Metcalf, "Book Review: *Power over People*," 796–97.

28. Young, *Power over People*, 69.

29. Young, *Power over People*, 17.

30. Young, *Power over People*, 188, 69.

31. Casper and Wellstone, *Powerline*, 3, 45.

32. Casper and Wellstone, *Powerline*, 286.

33. Casper and Wellstone, *Powerline*, 136, 151.

34. Casper and Wellstone, *Powerline*, 177.

35. Casper and Wellstone, *Powerline*, 201.

36. Casper and Wellstone, *Powerline*, 4, 287.

37. Dick Lowry, OHMS (TV movie), 1980.

38. The historian Bill Luckin said viewing OHMS inspired his own book-length study of public resistance to electric infrastructure, *Questions of Power*.

39. Casper and Wellstone, *Powerline*, 300.

40. Doyle, *Lines across the Land*, 144.

41. Hailey, *Overload*, 33, 159, 33, 91.

42. Hailey, *Overload*, 176, 191, 274.

43. Sovacool, *Handbook of Energy Security*, 30; Lovins and Lovins, *Brittle Power*, 131–32.

44. Hailey, *Overload*, 238.

45. Richard Friedman, "Power and Sex: Review of *Overload*," *New York Times*, January 28, 1979; Miles Beller, "Arthur Hailey Blows a Circuit," *Los Angeles Times*, February 11, 1979, N4.

46. Quotation about *Overload* and San Diego Gas and Electric appears in Beder, *Power Play*. See also Daniel Seligman, "They're Only Human," *Fortune Magazine*, February 12, 1979.

47. Thorough literature reviews of perception of transmission line research appear in Furby et al., "Public Perceptions," 19–43; Priestley and Deming-Beard, *Perception of Transmission Lines*; Doukas et al., "Electric Power Transmission," 979–88. The quotation about the failure of design appears in Tikalsky and Willyard, "Aesthetics and Public Perceptions," 38.

48. Joe C. Pohlman, "What *Is* the Public's Opinion on Transmission Towers and Poles?" *Electric Light and Power* (April 1973): 59–61.

49. Priestley and Evans, "Resident Perceptions of a Nearby Electric Transmission Line," 65–74.

50. Priestley and Deming-Beard, *Perception of Transmission Lines*.

51. Nelson et al., "Close and Connected," 1–30.

52. For scientific studies of EMFs, see Wertheimer and Leeper, "Electrical Wiring Configurations and Childhood Cancer," 273–84; and London et al., "Exposure to Residential Electric and Magnetic Fields and Risk of Childhood Leukemia," 923–37.

53. Linet et al., "Residential Exposure to Magnetic Fields and Acute Lymphoblastic Leukemia in Children," 1–7; Campion, "Power Lines, Cancer, and Fear," 44–46; Day, "Exposure to Power-Frequency Magnetic Fields and the Risk of Childhood Cancer," 1925–31; Kleinerman et al., "Are Children Living near High-Voltage Power Lines at Increased Risk of Acute Lymphoblastic Leukemia?" 512–15;

Kheifets et al., "Pooled Analysis of Extremely Low-Frequency Magnetic Fields and Childhood Brain Tumors," 752–61.

54. World Health Organization, "Extremely Low Frequency Fields," *Environmental Health Criteria* (Geneva, 2007), 238.

55. "Power Line Fears: Retro Report," *New York Times*, December 2, 2014, https://www.nytimes.com/video/health/100000003263065/power-line-fears.html.

56. Lindsay Carlton, "Can Power Lines Cause Cancer?" *Fox News Health*, January 3, 2017, http://www.foxnews.com/health/2017/01/03/can-living-near-power-lines-cause-cancer.html.

57. Patrick Devine-Wright has made significant arguments about the role of NIMBYism and transmission lines. His works include "Social Representations of Electricity Network Technologies," 357–73; "Explaining 'NIMBY' Objections to a Power Line," 761–81; "Making Electricity Networks 'Visible,'" 17–35; and "NIMBYism and Community Consultation in Electricity Transmission Network Planning," 115–28.

58. Canan Tasci, "Chino Hills Residents Protest Edison Towers at Shareholder Meeting," *San Gabriel Valley Tribune*, April 25, 2013, https://www.dailybulletin.com/2013/04/25/chino-hills-residents-protest-edison-towers-at-shareholder-meeting-2/.

59. Marianne Napoles, "Family Upset with Edison's Tactics," *Chino Champion / Chino Hills Champion*, May 18 2013, sec. A5.

60. "Alternate Proposed Decision of Commissioner Michael Peevey," California Public Utility Commission, July 11, 2013; "Decision Granting the City of Chino Hills' Petition for Modification of Decision 09-12-044 and Requiring Underground of Segment 8A of the Tehachapi Renewable Transmission Project."

61. Canan Tasci, "PUC President Celebrates Undergrounding in Chino Hills," *Inland Valley Daily Bulletin*, September 6, 2013, http://www.dailybulletin.com/2013/09/06/puc-president-celebrates-undergrounding-in-chino-hills/.

62. Interview with Bob Goodwin by the author, December 17, 2016, Chino Hills, California.

63. Interview with Goodwin by the author.

Conclusion

1. Edwards, "Infrastructure and Modernity," 185.

2. Megan Clark, "Aging U.S. Power Grid Blacks Out More than Any Other Developed Nation," *International Business Times*, July 17, 2014.

3. Bakke, *Grid*, xiv.

4. Massoud Amin, "The Smart-Grid Solution," *Nature*, July 11, 2013, 145.

5. Energy.gov, Office of Electricity Delivery and Energy Reliability, "Electric Disturbance Events," OE-417, for 2017, https://www.oe.netl.doe.gov/oe417.aspx.

6. Department of Energy, "Department of Energy, Top 9 Things You Didn't Know about America's Power Grid," Energy.gov, November 20, 2014, https://energy .gov/articles/top-9-things-you-didnt-know-about-americas-power-grid.

7. "Power Lines and Electrical Equipment Are a Leading Cause of California Wildfires," *LA Times*, October 17, 2017.

8. Michelle Davis and Steve Clemmer, "Power Failure: How Climate Change Puts Our Electricity at Risk—and What We Can Do," report by the Union of Concerned Scientists, April 2014, https://www.ucsusa.org/sites/default/files /legacy/assets/documents/Power-Failure-How-Climate-Change-Puts-Our -Electricity-at-Risk-and-What-We-Can-Do.pdf.

9. "How much of U.S. Carbon Dioxide Emissions Are Associated with Electricity Generation?" FAQ, U.S. Energy Information Administration, 2016, http://www .eia.gov/tools/faqs/faq.cfm?id=77&t=11.

10. See Zehner, *Green Illusions*.

11. Hong Yang, Xianjin Huang, Julian R. Thompson, "Tackle Pollution from Solar Panels," *Nature*, May 29, 2014.

12. Amin, "Smart-Grid Solution," 145.

13. As Benjamin K. Sovacool writes, "People will resist energy technologies that impede their freedom or appear to diminish their control." "Cultural Barriers," 371.

14. Lovins and Lovins, *Brittle Power*, 81–84.

15. Energy.gov, Office of Electricity Delivery and Energy Reliability, "Electric Disturbance Events," OE-417, for 2017, https://www.oe.netl.doe.gov/oe417.aspx.

16. "Panetta Warns of Dire Threat of Cyberattack on U.S.," *New York Times*, October 11, 2012.

17. Edison Electric Institute, "Transmission Projects at a Glance," December 2016, http://www.eei.org/issuesandpolicy/transmission/Pages/transmissionproject sat.aspx.

18. MacDonald et al., "Future Cost-Competitive Electricity Systems," 4–7.

19. Puneet Kollipara, "Better Power Lines Would Help U.S. Supercharge Renewable Energy, Study Suggests," *Science*, January 25, 2016.

20. Leichtman Research Group, "94.5 Million Get Broadband from Top Cable and Telephone Companies," 2017, http://www.leichtmanresearch.com/press /111617release.html.

21. Energy Information Administration, "Energy in Brief: How Many and What Kind of Power Plants Are There in the United States?" December 8, 2017, http://www.eia.gov/tools/faqs/faq.cfm?id=65&t=2.

22. Edison Electric Institute, "Transmission Facts," http://www.eei.org/issues andpolicy/transmission/Pages/default.aspx.

23. Vauhini Vara, "The Energy Interstate: A National System of Electricity Transmission Could Cut Power-Plant Emissions by 80 Percent," *Atlantic*, June 2016, https://www.theatlantic.com/magazine/archive/2016/06/the-energy-interstate/480756/.

24. Hall, *Out of Sight, Out of Mind Revisited*, 39.

25. Martin Kaste, "If Power Lines Fall, Why Don't They Go Underground?" NPR, February 1, 2012, https://www.npr.org/2012/02/01/146158822/if-power-lines-fall-why-dont-they-go-underground; Theodore J. Kury and the Conversation, "What Would It Take for the U.S. to Bury Its Power Lines?" *Fortune*, September 20, 2017, http://fortune.com/2017/09/19/hurricane-destruction-bury-power-lines; Darran Simon, "Isn't It Better to Just Bury Power Lines? That May Depend on Where You Live," CNN, September 14, 2017, https://www.cnn.com/2017/09/14/us/underground-power-lines-trnd/index.html.

26. Saint Consulting Group, "National Survey: Opposition to Power Transmission Lines," May 5, 2009, http://tscg.biz/files/transmission-line-survey-results-saint-consulting-1.pdf.

27. Michelle F. Bissonnette, Angela G. L. Piner, and Pamela J. Rasmussen, "Getting the Crop to Market: Siting and Permitting Transmission Lines on Buffalo Ridge, Minnesota," *Environment Concerns in Rights-of-Way Management: Eighth International Symposium*, 2008, https://doi.org/10.1016/B978-044453223-7.50030-7.

28. Most of the recent public perceptions research has been conducted in Europe. For recent studies of the range of public concerns of power lines in the United States, see International Electric Transmission Perception Project, *Perception of Transmission Lines*; Cain and Nelson, "What Drives Opposition"; and Nelson et al., "Close and Connected."

29. Cohen et al., "Empirical Analysis of Local Opposition to New Transmission Line."

30. Wuebben, "From Wire Evil to Power Line Poetics," 57–58.

31. Joseph H. Eto, "Building Electric Transmission Lines," Lawrence Berkeley National Laboratory, September 2016, https://eta.lbl.gov/sites/default/files/publications/lbnl-1006330.pdf.

32. Devine-Wright, "Making Electricity Networks 'Visible,'" 17–35.

33. Sovacool and Brown, "Compelling Tangle of Energy and American Society," 7.
34. Social science studies of energy include works by Bakke, Devine-Wright, Sovacool, and ethnographer, poet, and environmentalist Watts. The landmark text for energy humanities is Boyer and Szeman, *Energy Humanities*.
35. For recent interdisciplinary studies of "infrastructure," see Rubenstein et al., "Infrastructuralism," 575–86; Svensson, "From Optical Fiber to Conceptual Cyberinfrastructure"; and the international research group led by Lee Vinsel, "The Maintainers," themaintainers.org. *Cultural Anthropology* has devoted issues to "Infrastructure" (2011), https://culanth.org/curated_collections/11 -infrastructure; and "Anthropology Electric" (2015), https://culanth.org/articles /788-anthropology-electric.
36. Edwards, "Infrastructure and Modernity," 189.
37. Larkin, "Politics and Poetics of Infrastructure," 329.
38. 1970 National Power Survey, I-12-7.
39. "Code Words for Overhead Aluminum Electrical Conductors," Aluminum Association, http://www.aluminum.org/sites/default/files/Code%20Words %20for%20Overhead%20Aluminum%20Electrical%20Conductors.pdf.
40. Leatherbarrow, *Topographical Stories*, 1.
41. Samuel Wagstaff, "Talking with Tony Smith," *Art Forum* (December 1966).
42. Aldo Leopold, "What Is the University of Wisconsin Arboretum, Wild Life Refuge, and Forest Experiment Preserve?" Speech given at the dedication of University of Wisconsin Arboretum on June 17, 1934.
43. Heidegger, *Question Concerning Technology*, 295.

Bibliography

Adams, Edward Dean. *Niagara Power: The History of the Niagara Falls Power Company, 1886–1918.* 2 vols. New York: Bartlett Press, 1927.

Adams, Judith A. "The Promotion of New Technology through Fun and Spectacle: Electricity at the World's Columbian Exposition." *Journal of American Culture* 18, no. 2 (1995): 50.

American Landscape. No. 1. Engraved from Original and Accurate Drawings; Executed from Nature Expressly for This Work, and from Well Authenticated Pictures; With Historical and Topographical Illustrations. Edited and illustrated by A. B. Durand. Introduction by William Cullen Bryant. Additional illustrations by William James Bennet and Robert Walter Weir. New York: E. Bliss, 1830.

Armstrong, Neil. Foreword to *A Century of Innovation: Twenty Engineering Achievements That Transformed Our Lives*, edited by George Constable and Bob Somerville. Washington DC: Joseph Henry, 2003.

Ayers, Edward. *What Caused the Civil War? Reflections on the South and Southern History.* New York: Norton, 2005.

Badger, Reid. *The Great American Fair: The World's Columbian Exposition and American Culture.* Chicago: Nelson-Hall, 1979.

Bakke, Gretchen. *The Grid: The Fraying Wires between Americans and Our Energy Future.* New York: Bloomsbury, 2016.

Barrett, John Patrick. *Electricity at the Columbian Exposition.* Chicago: R. R. Donnelly & Sons, 1894.

Bazerman, Charles. *The Languages of Edison's Light.* Cambridge MA: MIT University Press, 1999.

Beder, Sharon. *Power Play: The Fight to Control the Worlds Electricity.* New York: New Press, 2003.

Bellamy, Edward. *Looking Backward: 2000–1887.* 1888. Reprinted with an introduction by Walter James Miller. New York: Signet Classic, 2000.

Bellour, Raymond. "To Alternate/Narrate (on *The Lonedale Operator*)." In *The Analysis of Film*, edited by Constance Penley, 262–78. Bloomington, Indiana University Press, 2000.

Bennett, E. H. "Aesthetic Considerations Affecting Power Development." *Annals of the American Academy of Political and Social Science* 118 (March 1925): 116–19.

Benson, Etienne. "Generating Infrastructural Invisibility: Insulation, Interconnection, and Avian Excrement in the Southern California Power Grid." *Environmental Humanities* 6 (2015): 103–30.

Berte, Leigh Ann Litwiller. "Mapping 'The Octopus': Frank Norris' Naturalist Geography." *American Literary Realism* 37, no. 3 (Spring 2005): 202–24.

Billington, David P., and David P. Billington Jr. *Power, Speed, and Form: Engineers and the Making of the Twentieth Century*. Princeton: Princeton University Press, 2006.

Blondheim, Menaham. "Three Phases in the Diffusion and Perception of American Telegraphy." *Technology, Pessimism and Postmodernism*, 77–92. Dordrecht, Netherlands: Springer, 1994.

——. *News over the Wires: The Telegraph and the Flow of Public Information in America, 1844–1897*. Cambridge: Harvard University Press, 1994.

Bogue, Allan G. *Frederick Jackson Turner: Strange Roads Going Down*. Norman: University of Oklahoma Press, 1998.

Bonneville Power Administration (BPA). *Hydro: Power to Make the American Dream Come True*. Film. Portland OR: BPA Motion Picture Information Division, 1939.

Bottemiller, Steven C., and Marvin L. Wolverton. "The Price Effects of HVTLs on Abutting Homes." *Appraisal Journal* (Winter 2013): 45–62.

Bowser, Eileen, ed. *Biograph Bulletins, 1908–1912*. New York: Octagon Books, 1973.

Boyer, Dominic, and Imre Szeman, eds. *Energy Humanities: An Anthology*. Baltimore: Johns Hopkins University Press, 2017.

Brigham, Jay. *Empowering the West: Electrical Politics before FDR*. Lawrence: University Press of Kansas, 1988.

Brown, Charles. *First Telegraph Line across the Continent: Charles Brown's 1861 Diary*. Edited by Dennis N. Mihelich and James E. Potter. Lincoln: Nebraska State Historical Society Books, 2011.

Bryant, William Cullen. *The Life and Works of William Cullen Bryant*. Vol. 2: *Poems*. Edited by Parke Goodwin. New York: D. Appleton, 1889.

——. *The Life and Works of William Cullen Bryant*. Vol. 5: *Prose Writings*. Edited by Parke Goodwin. New York: D. Appleton, 1889.

Cable, Mary. *The Blizzard of '88*. New York: Athenaeum, 1988.

Cain, Nicholas, and Hal Nelson. "What Drives Opposition to High-Voltage Transmission Lines?" *Land Use Policy* 33 (2013): 204–13.

Campion, Edward W. "Power Lines, Cancer, and Fear." *New England Journal of Medicine* 337 (1997): 44–46.

Carlson, Bernard. *Tesla: Inventor of the Electric Age*. Princeton: Princeton University Press, 2014.

Carson, Rachel. *Silent Spring: 40th Anniversary Edition*. New York: Houghton Mifflin, 2002.

Casper, Barry M., and Paul Wellstone. *Powerline: The First Battle of America's Energy War*. 1981. Reprint, Minneapolis: University of Minnesota Press, 2003.

Casson, Herbert Newton. *The History of the Telephone*. Chicago: A. C. McClurg, 1910.

Cikovsky, Nicolai, Jr. Introduction to *Lectures on the Affinity of Painting with the Other Fine Arts* by Samuel F. B. Morse. Columbia: University of Missouri Press, 1983.

Coe, Lewis. *The Telegraph: A History of Morse's Invention and Its Predecessors in the United States*. Jefferson NC: McFarland, 2003.

Cohen, Jed, et al. "An Empirical Analysis of Local Opposition to New Transmission Lines across the EU-27." *Energy Journal, International Association for Energy Economics* 37, no. 3 (2016): 59–82.

Cohn, Julie A. *The Grid: Biography of an American Technology*. Cambridge MA: MIT Press, 2017.

Cole, Thomas. "Essay on American Scenery." *American Monthly Magazine* 1 (1836): 1–12. Reprinted in *Thomas Cole: The Collected Essays and Prose Sketches*, edited by Marshall Tymn, 3–19. St. Paul: John Colet Press, 1980.

Conron, John, ed. *The American Landscape: A Critical Anthology of Prose and Poetry*. New York: Oxford University Press, 1973.

Cronon, William. *Changes in the Land: Indians, Colonists, and the Ecology of New England*. New York: Hill & Wang, 1983.

———. *Nature's Metropolis: Chicago and the Great West*. New York: Norton, 1992.

Czitrom, Daniel J. *Media and the American Mind: From Morse to McLuhan*. Chapel Hill: University of North Carolina Press, 1983.

Cunningham, Joseph J. *New York Power*. North Charleston SC: CreateSpace, 2013.

Day, Nick. "Exposure to Power-Frequency Magnetic Fields and the Risk of Childhood Cancer." *Lancet* 354 (1999): 1925–31.

Delbourgo, James. *A Most Amazing Scene of Wonders: Electricity and Enlightenment in Early America*. Cambridge MA: Harvard University Press, 2006.

Devine-Wright, Patrick. "Explaining 'NIMBY' Objections to a Power Line: The Role of Personal, Place Attachment and Project-Related Factors." *Environment and Behavior* 45 (2013): 761–81.

_____. "Making Electricity Networks 'Visible': Industry Actor Constructions of 'Publics' and Public Engagement in Infrastructure Planning." *Public Understanding of Science* 21 (2012): 17–35.

_____. "NIMBYism and Community Consultation in Electricity Transmission Network Planning." In *Renewable Energy and the Public: From NIMBY to Participation*, edited by Patrick Devine-Wright, 115–58. London: Earthscan, 2011.

_____. "Social Representations of Electricity Network Technologies: Exploring Processes of Anchoring and Objectification through the Use of Visual Research Methods." *British Journal of Social Psychology* 48 (June 2009): 357–73.

Dods, John Bovee. *The Philosophy of Electrical Psychology: In a Course of Twelve Lectures*. New York: Fowler & Wells, 1888.

Doukas, Haris, et al. "Electric Power Transmission: An Overview of Associated Burdens." *International Journal of Energy Research* 35, no. 11 (2011): 979–88.

Doyle, Jack. *Lines across the Land: Rural Electric Cooperatives: The Changing Political of Energy in Rural America*. Edited by Vic Reinemer. Washington DC: Environmental Policy Institute, 1979.

Drew, Alan. "1901 Transmission Line: The Great Carquinez Straits Crossing." *IEEE Power and Energy* 8, no. 3 (May–June 2010): 80–88.

Dreyfuss, Henry, and Associates. *Electric Transmission Structures: A Design Research Program*. Prepared for the Electric Research Council's Esthetic Designs for Transmission Structures research project. New York: Edison Electric Institute, 1968.

Durand, Asher B. "Letters on Landscape Painting." *The Crayon: A Journal Devoted to Graphic Arts, and the Literature Related to Them* (1855). Reprinted in *Kindred Spirits: Asher B. Durand and the American Landscape*, edited by Linda Ferber, 231–51. New York: Brooklyn Museum of Art in association with D. Giles, London, 2007.

Edwards, Jonathan. *A Jonathan Edwards Reader*. Edited by John E. Smith, Harry S. Stout, and Kenneth P. Minkea. New Haven: Yale University Press, 1995.

Edwards, Paul. "Infrastructure and Modernity: Force, Time, and Social Organization in the History of Sociotechnical Systems." In *Modernity and Technology*, edited by Thomas J. Misa, Philip Brey, and Andrew Feenberg. Cambridge MA: MIT Press, 2003.

"Electric Power Transmissions of Southern California." *Journal of Electricity, Power and Gas* 9, no. 3 (March 1900): 45–47.

Elliot, Peter, and David Wadley. "Coming to Terms with Power Lines." *International Planning Studies* 17, no. 2 (2012): 179–201.

Emerson, Ralph Waldo. *Essays and Poems*. Edited by Joel Porte. New York: Library of America, 1983.

——. *The Journals and Miscellaneous Notebooks*. Edited by William H. Gilman and Ralph H. Orth et al. 16 vols. Cambridge: Belknap Press of Harvard University Press, 1960–82.

——. "Poetry and Imagination." In *Ralph Waldo Emerson*, edited by Richard Poirier. Oxford: Oxford University Press, 1966.

Essig, Mark. *Edison and the Electric Chair: A Story of Light and Death*. New York: Walker, 2003.

"An Evening with the Telegraph Wires." *Atlantic Monthly* 2, no. 11 (September 1858): 489–95.

Faragher, John Mack. "Introduction: A Nation Thrown Back upon Itself." In *Rereading Frederick Jackson Turner: "The Significance of the Frontier in American History," and Other Essays*, 1–10. New Haven: Yale University Press, 1998.

Ferber, Linda S. *Kindred Spirits: Asher B. Durand and the American Landscape*. New York: Brooklyn Museum of Art in association with D. Giles, London, 2007.

Flinchum, Russel. *Henry Dreyfuss, Industrial Designer: The Man in the Brown Suit*. New York: Rizzoli in conjunction with Cooper-Hewitt National Design Museum, 1997.

French, Daniel. *When They Hid the Fire: A History of Electricity and Invisible Energy in America*. Pittsburgh: University of Pittsburgh Press, 2017.

Furby, Linda, et al. "Public Perceptions of Electric Power Transmission Lines." *Journal of Environmental Psychology* 8, no. 1 (1988): 19–43.

Gamble, James. "Early Reminiscences of the Telegraph on the Pacific Coast." *Californian Magazine* (1881). http://www.telegraph-history.org/transcontinental-telegraph/index.html.

——. "Wiring a Continent: The Making of the U.S. Transcontinental Telegraph Line." *Californian Magazine* (1881). http://www.telegraph-history.org/transcontinental-telegraph/index.html.

Gillespie, Sarah Kate. *The Early American Daguerreotype: Cross-Currents in Art and Technology*. Cambridge MA: MIT Press in association with Lemelson Center, 2015.

Gilmore, Paul. *Aesthetic Materialism: Electricity and American Romanticism*. Stanford: Stanford University Press, 2008.

Glaser, Leah. *Electrifying the Rural American West: Stories of Power, People, and Place*. Lincoln: University of Nebraska Press, 2009.

————. "Nice Towers, eh? Evaluating a Transmission Line in Arizona." CRM: *Cultural Resource Management* (U.S. Department of the Interior, National Park Service) 20, no. 14 (1997): 23–24.

Goulty, George A. *Visual Amenity Aspects of High Voltage Transmission*. New York: Wiley, 1990.

Griffith, D. W. *The Girl and Her Trust*. Biograph Studios, *Los Angeles*, 1912. Screenplay by George Hennessy. Cast featuring Dorothy Bernard (Grace), Wilfred Lucas (Jack).

————. *The Lonedale Operator*. Biograph Studios, 1911.

————. *The Man Who Invented Hollywood: The Autobiography of D. W. Griffith*. Edited and annotated by James Hart. Louisville KY: Touchstone Publishing, 1972.

Gunning, Tom. "Heard over the Phone: The Lonely Villa and the de Lorde Tradition of the Terrors of Technology." *Screen* 32, no. 2 (Summer 1991): 184–96.

————. "Systematizing the Electric Message: Narrative Form, Gender, and Modernity in *The Lonedale Operator*." In *American Cinema's Transitional Era: Audiences, Institutions, Practices*, edited by Charlie Keil and Shelley Stamp, 15–50. Berkeley: University of California Press, 2004.

Guthrie, Woodie. *From California to New York Island*. New York: Guthrie Children's Trust Fund, 1960.

Hailey, Arthur. *Overload*. New York: Doubleday & Co., 1979.

Hall, Kenneth L. *Out of Sight, Out of Mind Revisited: An Updated Study on the Undergrounding of Overhead Power Lines*. Washington DC: Edison Electric Institute, 2012.

Halliday, Sam. *Science and Technology in the Age of Hawthorne, Melville, Twain, and James: Thinking and Writing Electricity*. New York: Palgrave Macmillan, 2007.

Halliwell, Leslie. *Halliwell's Who's Who in the Movies*. Edited by John Walker. New York: HarperCollins, 1999.

Harlow, Alvin F. *Old Wires and New Waves: The History of the Telegraph, Telephone, and Wireless*. New York: D. Appleton-Century, 1936.

Hawthorne, Nathaniel. *The House of the Seven Gables*. 1851. Edited by Robert S. Levine. New York: Norton, 2005.

Heidegger, Martin. *The Question Concerning Technology, and Other Essays*. New York: Harper, 1977.

Hietala, Thomas R. *Manifest Design: American Exceptionalism and Empire*. Ithaca: Cornell University Press, 2003.

Hochfelder, David. *The Telegraph in American, 1832–1920*. Baltimore: Johns Hopkins University Press, 2012.

Holmes, Oliver Wendell. *The Writings of Oliver Wendell Holmes*. Cambridge: Riverside Press, 1891.

Holmes, Richard. *The Age of Wonder: How the Romantic Generation Discovered the Beauty and Terror of Science*. New York: Vintage Books, 2008.

Howe, David Walker. *What Hath God Wrought: The Transformation of America, 1815–1848*. New York: Oxford University Press, 2007.

Hughes, Thomas P. "The Evolution of Large Technological Systems." In *The Social Construction of Technological Systems: New Directions in the Sociology and History of Technology*, edited by Wiebe E. Bijker, Thomas P. Hughes, and Trevor Pinch, 51–82. Cambridge MA: MIT Press 2012.

———. *The Human Built World: How to Think about Technology and Culture*. Chicago: University of Chicago Press, 2005.

———. *Networks of Power: Electrification in Western Society, 1880–1930*. Softshell Books ed. Baltimore: Johns Hopkins University Press, 1993.

———. "Technological Momentum." In *Does Technology Drive History? The Dilemma of Technological Determinism*, edited by Merritt Roe Smith and Leo Marx, 101–14. Cambridge MA: MIT Press, 1994.

Hunter, Louis C., and Lynwood Bryant. *A History of Industrial Power in the United States, 1780–1930*, vol. 3: *The Transmission of Power*. Charlottesville: University Press of Virginia, 1991.

Huurdeman, Anton E. *The Worldwide History of Telecommunications*. Hoboken NJ: John Wiley & Sons, 2003.

Irwin, William. *The New Niagara: Tourism, Technology, and the Landscape of Niagara Falls, 1776–1917*. University Park: Penn State University Press, 1996.

James, William. *The Complete Works of William James: The Principles of Psychology*. Cambridge: Harvard University Press, 1981.

———. *The Correspondence of William James*. Edited by Ignas K. Skrupskelis and Elizabeth M. Berkeley. Vol. 2. Charlottesville: University Press of Virginia, 1993.

Jefferson, Thomas. *The Life and Selected Writings of Thomas Jefferson*. Edited by Adrienne Koch and William Peden. New York: Random House, 1998.

Jepsen, Thomas P. "The Telegraph Comes to Colorado: A New Technology and Its Consequences." *Essays and Monographs in Colorado History* 7 (1987): 1–23.

John, Richard R. *Network Nation: Inventing American Telecommunications*. Cambridge: Harvard University Press, 2010.

Johnson, Rochelle. *Passions for Nature: Nineteenth-Century America's Aesthetics of Alienation*. Athens: University of Georgia Press, 2009.

Johnson, Rossiter, ed. *A History of the World's Columbian Exposition Held in Chicago in 1893*. New York: D. Appleton, 1897–98.

Jones, Alexander. *Historical Sketch of the Electric Telegraph: Including Its Rise and Progress in the United States*. New York: George P. Putnam, 1852.

Jones, Christopher. *Routes of Power: Energy and Modern America*. Cambridge: Harvard University Press, 2014.

Jonnes, Jill. *Empires of Light: Edison, Tesla, Westinghouse, and the Race to Electrify the World*. New York: Random House, 2003.

Joyce, James. *A Portrait of the Artist as a Young Man*. 1916. Reprint, New York: B. W. Huebsch, 1921.

Karady, George. "Environmental Impact of Transmission Lines." In *Electric Power Generation, Transmission, and Distribution*, edited by Leonard L. Grigsby, 20:1–20. Boca Raton FL: CRC Press, Taylor Francis, 2012.

Kheifets, Leeka, et al. "A Pooled Analysis of Extremely Low-Frequency Magnetic Fields and Childhood Brain Tumors." *American Journal of Epidemiology* 172, no. 7 (2010): 752–61.

Kleinerman, Ruth A., et al. "Are Children Living Near High-Voltage Power Lines at Increased Risk of Acute Lymphoblastic Leukemia?" *American Journal of Epidemiology* 151, no. 5 (2000): 512–15.

Kline, Ronald R. *Consumers in the Country: Technology and Social Change in Rural America*. Baltimore: Johns Hopkins University Press, 2002.

Knight, Arthur. *The Liveliest Art: A Panoramic History of the Movies*. New York: New American Library, 1957.

Larkin, Brian. "The Politics and Poetics of Infrastructure." *Annual Review of Anthropology* 42 (September 2013): 327–43.

Larkin, Oliver. *Samuel F. B. Morse and American Democratic Art*. Boston: Little, Brown, 1954.

Leatherbarrow, David. *Topographical Stories: Studies in Landscape and Architecture*. Philadelphia: University of Pennsylvania Press, 2015.

Leopold, Aldo. "Conservation Esthetic." *Bird-Lore* 40 (1938): 109.

———. "What Is the University of Wisconsin Arboretum, Wild Life Refuge, and Forest Experiment Preserve?" Speech given at the dedication of University of Wisconsin Arboretum on June 17, 1934.

Levy, Eugene. "The Aesthetics of Power: High Voltage Transmission Systems and the American Landscape." *Technology and Culture* 38, no 3. (July 1997): 575–607.

Lieberman, Jennifer L. *Power Lines: Electricity in American Life and Letters, 1882–1952*. Cambridge MA: MIT Press, 2017.

Linet, Martha S., et al. "Residential Exposure to Magnetic Fields and Acute Lympho-blastic Leukemia in Children." *New England Journal of Medicine* 337 (1997): 1–7.

London, Stephanie, et al. "Exposure to Residential Electric and Magnetic Fields and Risk of Childhood Leukemia." *American Journal of Epidemiology* 134, no. 9, (1991): 923–37.

Loss, Scott R., Tom Will, and Peter P. Marra. "Refining Estimates of Bird Collision and Electrocution Mortality at Power Lines in the United States." *PLOS ONE* 9, no. 7 (2014): e101565.

Lovins, Amory B., and L. Hunter Lovins. *Brittle Power: Energy Strategy for National Security*. 1982. Reprint, Andover MA: Brick House Publishing, 2001.

Low, George P. "The World's Longest Electric Power Transmission." *Journal of Electricity, Power, and Gas* 11, no. 7 (July 1901): 145–73.

Luckin, Bill. *Questions of Power: Electricity and Environment in Inter-War Britain*. Manchester: Manchester University Press, 1990.

Lusted, David. *The Western: Inside Film*. New York: Longman, 2003.

MacDonald, Alexander, et al. "Future Cost-Competitive Electricity Systems and Their Impact on U.S. CO_2 Emissions." *Nature Climate Change* 6 (2016): 526–31.

MacDougall, Robert. "The Wire Devils: Pulp Thrillers, the Telephone, and Action at a Distance in the Wiring of a Nation." *American Quarterly* 58, no. 3 (September 2006): 71–74.

Marvin, Carolyn. *When Old Technologies Were New: Thinking about Electric Communication in the Late Nineteenth Century*. New York: Oxford University Press, 1988.

Marx, Leo. *The Machine in the Garden: Technology and the Pastoral Ideal in America*. New York: Oxford University Press, 1967.

Melville, Herman. *Moby-Dick, or, The Whale*. 1851. Edited by Hershel Parker and Harrison Hayford. New York: Norton, 1976.

Menke, Richard. "Telegraphic Realism: Henry James's *In the Cage*." *PMLA* 115 (2000): 975–90.

———. *Telegraphic Realism: Victorian Fiction and Other Information Systems*. Stanford: Stanford University Press, 2008.

Metcalf, Lee. "Book Review: *Power over People*." *Political Research Quarterly* 26, no. 4 (1973): 796–97. https://doi.org/10.1177/106591297302600418.

Miller, Charles A. *Jefferson and Nature: An Interpretation*. Baltimore: Johns Hopkins University Press, 1988.

Miller, Perry. *Nature's Nation*. Cambridge: Belknap Press of Harvard University Press, 1967.

Mitchell, W. J. T., ed. *Landscape and Power*. 2nd ed. Chicago: University of Chicago Press, 2002.

Morley, Erica L., and Daniel Robert. "Electric Fields Elicit Ballooning in Spiders." *Current Biology* 28, no. 14, July 23, 2018, 2324–30.

Morse, Samuel F. B. "'Academies of the Arts: A Discourse.' Delivered on Thursday May 3, 1827, in the Chapel of Columbia College, before the National Academy of Design on Its First Anniversary." New York: G. & C. Carvill, 1827. Reprinted in *North American Review* 26, no. 58 (January 1828): 207–24.

———. *Lectures on the Affinity of Painting with the Other Fine Arts*. Edited and with an introduction by Nicolai Cikovsky Jr. Columbia: University of Missouri Press, 1983.

———. *Letters and Journals*. 2 vols. Edited and supplemented by Edward Lind Morse. Boston: Houghton Mifflin, 1914.

Mumford, Lewis. *Technics and Civilization*. 1934. Reprint, Chicago: University of Chicago Press, 2010.

Myers, William A. *Iron Men and Copper Wires: A Centennial History of Southern California Edison Company*. Glendale CA: Trans-Anglo Books, 1983.

Nabokov, Vladimir. *The Stories of Vladimir Nabokov*. New York: Vintage, 1997.

Naremore, James. *Orson Welles's Citizen Kane: A Casebook*. Oxford: Oxford University Press, 2004.

Nash, Roderick. *Wilderness and the American Mind*. 4th ed. 1967. Reprint, New Haven: Yale University Press, 2001.

National Rural Electric Cooperative Association. *The Next Greatest Thing: Fifty Years of Rural Electrification in America*. Washington DC: National Rural Electric Cooperative Association, Public & Association Affairs Department, 1984.

Nelson, Hal T., et al. "Close and Connected: The Effects of Proximity and Social Ties on Citizen Opposition to Electricity Transmission Lines." *Environment and Behavior* (May 2017): 1–30.

Norris, George. *Fighting Liberal: The Autobiography of George W. Norris*. 1945. Reprint, Lincoln: Bison Books, University of Nebraska Press, 1972.

Novak, Barbara. *Nature and Culture: American Landscape and Painting, 1825–1875*. Rev. ed., with new preface. New York: Oxford University Press, 1995.

Nye, David E. *American Technological Sublime*. Cambridge MA: MIT Press, 1994.

———. *Consuming Power: A Social History of American Energies*. Cambridge MA: MIT Press, 1997.

———. *Electrifying America: Social Meanings of a New Technology, 1880–1940*. Cambridge MA: MIT Press, 1992.

Olsen, Charles. *Collected Prose*. Berkeley: University of California Press, 1997.

O'Neill, John J. *Prodigal Genius: The Life of Nikola Tesla*. New York: Cosimo, 2006.

O'Sullivan, John. "Annexation." *United States Magazine and Democratic Review* (New York) 17 (1845): 5–6, 9–10.

Otis, Laura. *Networking: Communicating with Bodies and Machines in the Nineteenth Century*. Ann Arbor: University of Michigan Press, 2001.

Pomeroy, Earl. "Toward a Reorientation of Western History: Continuity and Environment." *Mississippi Valley Historical Review* 41, no. 4 (March 1955): 579–600.

Porsius, Jarry, et al. "'They Give You Lots of Information, but Ignore What It's Really About': Residents' Experiences with the Planned Introduction of a New High-Voltage Power Line." *Journal of Environmental Planning and Management* 59, no. 8 (October 2015): 1–18.

Priestley, Tom. *Perceived Effects of Electric Transmission Facilities: A Review of Survey-Based Studies*. Prepared for the Siting and Environmental Planning Task Force. Washington DC: Edison Electric Institute, 1992.

Priestley, Tom, and Mary Deming-Beard. *Perception of Transmission Lines: Summary of Surveys and Framework for Further Research*. International Electric Transmission Perception Project. Washington DC: Edison Electric Institute, 1996.

Priestley, Tom, and Gary Evans. "Resident Perceptions of a Nearby Electric Transmission Line." *Journal of Environmental Psychology* 16, no. 1 (1996): 65–74.

Purdon, James. "Electric Cinema, Pylon Poetry." *Amodern 2: Network Archaeology* (October 2013). http://amodern.net/article/electric-cinema-pylon-poetry/.

Pursell, Carroll. *The Machine in America: A Social History of Technology*. Baltimore: Johns Hopkins University Press, 1995.

Reid, Badger. *The Great American Fair: The World's Columbian Exposition and American Culture*. Chicago: Nelson-Hall, 1979.

Richardson, Robert D. *William James: In the Maelstrom of American Modernism*. Boston: Houghton Mifflin, 2006.

Robinson, Charles Mulford. *Better Binghamton: A Report to the Mercantile Press Club of Binghamton*. [Cleveland OH: printed by J. B. Savage Co.], 1911.

———. *A City Plan for Raleigh: A Report to the Civic Department of the Woman's Club of Raleigh*. [Raleigh NC: printed by the Woman's Club of Raleigh], 1913.

Rubenstein, Michael, Bruce Robbins, and Sophia Beal. "Infrastructuralism: An Introduction." *MFS: Modern Fiction Studies* 61, no. 4 (2015): 575–86. Russell, Edmund, et al. "The Nature of Power: Synthesizing the History of Technology and Environmental History." *Technology and Culture* 52, no. 2 (2011): 246–59.

Schatzberg, Eric. "Culture and Technology in the City: Opposition to Mechanized Street Transportation in Late-Nineteenth-Century America." In *Technology of*

Power, edited by Michael Thad Allen and Gabrielle Hecht, 57–92. Cambridge MA: MIT Press, 2001.

Schickel, Richard. *D. W. Griffith: An American Life*. New York: Simon & Schuster, 1984.

Schivelbusch, Wolfgang. *The Railway Journey: Industrialization of Time and Space in the Nineteenth Century*. Berkeley: University of California Press, 1986.

Seifer, Marc. *Wizard: The Life and Times of Nikola Tesla, Biography of a Genius*. New York: Citadel Press, 1998.

Shaffner, Taliaferro Preston. *Telegraph Companion: Devoted to the Science and Art of the Morse American Telegraph*, vols. 1–2. New York: Pudney & Russell, 1854–55.

Silverman, Kenneth. *Lightning Man: The Accursed Life of Samuel F. B. Morse*. New York: Da Capo Press, 2004.

Simon, Linda. *Dark Light: Electricity and Anxiety from the Telegraph to the X-Ray*. Orlando: Harcourt, 2004.

Skrupskelis, Ignas K., and Elizabeth M. Berkeley, eds. *The Correspondence of William James*, vol. 2. Charlottesville: University Press of Virginia, 1993.

Slotkin, Richard. *Gunfighter Nation: The Myth of the Frontier in Twentieth-Century America*. Norman: University of Oklahoma Press, 1998.

Smith, Michael L. "Recourse of Empire: Landscapes of Progress in Technological America." In *Does Technology Drive History? The Dilemma of Technological Determinism*, edited by Merritt Roe Smith and Leo Marx, 37–52. Cambridge MA: MIT Press, 1994.

Sovacool, Benjamin. "The Cultural Barriers to Renewable Energy and Energy Efficiency in the United States." *Technology in Society* 31 (2009): 365–73.

———, ed. *The Routledge Handbook of Energy Security*. London: Routledge, 2011.

Sovacool, Benjamin, and Marilyn A. Brown. "The Compelling Tangle of Energy and American Society." In *Energy and American Society: Thirteen Myths*, 1–21. Dordrecht, Netherlands: Springer, 2007.

Staiti, Paul J. *Samuel F. B. Morse*. Cambridge Monographs on American Artists. Cambridge: Cambridge University Press, 1990.

Starosielski, Nicole. *Undersea Network*. Durham NC: Duke University Press, 2015.

Stringer, Arthur. *Phantom Wires*. 1907. Supertales of Modern Mystery Series. New York: McKinlay, Stone & Mackenzie, 1923.

———. *The Wire Tappers*. 1906. Supertales of Modern Mystery Series. New York: McKinlay, Stone & Mackenzie, 1922.

Sullivan, Joseph P. "Fearing Electricity: Overhead Wire Panic in New York City." *IEEE Technology and Society Magazine* 14, no. 3 (1995): 8–16.

Svensson, Patrik. "From Optical Fiber to Conceptual Cyberinfrastructure." *Digital Humanities Quarterly* 5, no. 1 (2011). http://www.digitalhumanities.org/dhq/vol /5/1/000090/000090.html.

Tesla, Nikola. "A New System of Alternate Current Systems and Motors." New York Lecture, May 16, 1888. Reprinted in *The Inventions, Research, and Writings of Nikola Tesla*. 1893. Reprint, New York: Barnes & Noble Books, 1992. 7–25.

———. "On Electricity." Buffalo Speech, January 11, 1897. The transcript first appeared in the *Electrical Review*, January 27, 1897. Reprinted in *Nikola Tesla: Lectures, Patents, Articles*. Compiled by Vojin Popović, Radoslav Horvat, and Nikola Nikolić, 101–8. Belgrade, Yugoslavia: Nikola Tesla Museum, 1956.

———. "On the Problem of Increasing Human Energy: With Special Reference to Harnessing of the Sun's Energy." *Century Magazine* (1900). Reprinted as *The Problem of Increasing Human Energy*. Belgrade, Serbia: Nikola Tesla Museum, 2000.

Thompson, Kristin. *The Lonedale Operator*. Griffith Project, vol. 5: *Films Produced in 1911*, edited by Paolo Cherchi Usai, 18–20. London: British Film Institute, 2001.

Thompson, Robert Luther. *Wiring a Continent: The History of the Telegraph Industry in the United States, 1832–1866*. Princeton: Princeton University Press, 1947.

Thoreau, Henry David. *The Journal of Henry David Thoreau*. 14 vols. Edited by Bradford Torrey and Francis H. Allen. Boston: Houghton Mifflin, 1906.

———. *A Week on the Concord and Merrimack Rivers, Walden, The Maine Woods, Cape Cod*. Edited by Robert Sayre. New York: Library of America, 1989.

Tikalsky, Susan M., and C. J. Willyard. "Aesthetics and Public Perception of Transmission Structures: A Brief History of the Research." *Right of Way* (Electric Power Research Institute [EPRI], Palo Alto CA) (March–April 2007): 28–32.

Turnbull, Laurence. *The Electro Magnetic Telegraph: With an Historical Account of Its Rise, Progress, and Present Condition*. Philadelphia: A. Hart, 1853.

Turner, Frederick Jackson. "The Significance of the Frontier in American History." 1893. In *Rereading Frederick Jackson Turner: "The Significance of the Frontier in American History," and Other Essays*, edited by John Mack Faragher. New Haven: Yale University Press, 1998.

U.S. Bureau of the Census. *Agriculture: 1950*. Vol. 5: *Special Reports*, pt. 6, *Agriculture 1950—A Graphic Summary*. Washington DC: U.S. Government Printing Office, 1952.

———. *Historical Statistics of the United States, Colonial Times to 1857*. Washington DC: Government Printing Office, 1959.

———. *Historical Statistics of the United States: Colonial Times to 1970: Bicentennial Edition*. Washington DC: Government Printing Office, 1976.

———. *Population by Counties: 1890, 1900, 1910*. Washington DC: Government Printing Office, 1912.

———. *Telephones and Telegraphs: 1902*. Edited by Thomas Commerford Martin, Arthur Vaughan Abbott, and William Mayer. Washington DC: Government Printing Office, 1906.

Vincent, W. G., Jr. "The Interconnected Transmission System of California." *Journal of Electricity* 54, no. 12, June 15, 1925, 563–75.

Ware, Eugene F. *The Indian War of 1864*. Lincoln: University of Nebraska Press, 1960.

Wasik, John. *The Merchant of Power: Sam Insull, Thomas Edison, and the Creation of the Modern Metropolis*. New York: St. Martin's Press, 2015.

Weitenkampf, Frank. *Manhattan Kaleidoscope*. New York: Charles Scribner's Sons, 1947.

Wertheimer, Nancy, and Ed Leeper. "Electrical Wiring Configurations and Childhood Cancer." *American Journal of Epidemiology* 109, no. 3 (April 1979): 273–84.

White, Trumbull, and William Ingleheart, eds. *The World's Columbian Exposition, Chicago, 1893*. Chicago: J. K. Hastings, 1983.

Whitman, Walt. *Complete Poetry and Collected Prose*. New York: Library of America, 1982.

Williams, James C. *Energy and the Making of Modern California*. Akron: University of Akron Press, 1997.

Watts, Laura. *Energy at the End of the World: An Orkney Islands Saga*. Cambridge MA: MIT Press, 2018.

Wilson, Eric. *Emerson's Sublime Science*. New York: St. Martin's Press, 1999.

Worster, Donald. *Under Western Skies: Nature and History in the American West*. New York: Oxford University Press, 1992.

Wuebben, Daniel. "From Wire Evil to Power Line Poetics: The Ethics and Aesthetics of Renewable Transmission." *Energy Research & Social Science* 30 (August 2017): 53–60.

Yates, JoAnne. "The Telegraph's Effect on Nineteenth Century Markets and Firms." *Business and Economic History* 15 (1986): 149–63.

Young, Louise B. *Power over People*. 1973. Updated ed. New York: Oxford University Press, 1992.

Young, Paul. "Media on Display: A Telegraphic History of Early American Cinema." In *New Media, 1740–1915*, edited by Lisa Gitelman and Geoffrey B. Pingree, 229–64. Cambridge MA: MIT Press, 2003.

Zehner, Ozzie. *Green Illusions: The Dirty Secrets of Clean Energy and the Future of Environmentalism*. Lincoln: University of Nebraska Press, 2012.

Index

Thoreau; Henry David, 18, 20, 176; and landscape, 22–23; and telegraph harp, xx, 41–45
transmission lines, xvii, xix, 4, 6–7, 11–13, 49–50, 55–56, 61, 67, 89–90, 93, 99–100, 109–13, 131–32, 135, 138, 139, 142–43, 146–47, 150–51, 156, 159–61, 166–67, 169–71, 176, 178–79; aesthetic designs for, 118–27, 129; miles of, 4, 112, 141, 176; poetics of, 183–90; towers and pylons, 109–13, 121–22. *See also* power line battles
tubular steel poles (TSPS), xxi–xxiii, 119
Turnbull, Laurence, 35–37
Turner, Frederick Jackson: frontier thesis, 49, 52–56, 109

undergrounding (of electric lines), xvii–xviii, xxii, 3–4, 13, 57, 91, 124–25, 143, 161–63, 176–78; in New York, 69–75

Vail, Alfred, 1, 2, 3
visual salience, xviii–xx, 78, 105–6, 125–26, 146, 156–57, 161, 181–84, 189–90

Walden; or, Life in the Woods (Thoreau), xx, 24, 41–45

Welles, Orson, 134, 136
Wells, H. G., 66–67
Wellstone, Paul. See *Powerline: The First Battle in America's Energy War*
western tropes, 54, 87, 91–97
Western Union, 2, 17, 29, 67, 68–69, 71, 81, 96
Westinghouse, George, 57, 63–64, 73, 88–89
"white coal." *See* hydroelectric power
Whitman, Walt, 18, 20, 137, 146; "I Sing the Body Electric," 23–24
wilderness, 7, 9, 52–54
wireless, xvii, 13, 60–61, 83–84, 127
wires, xiii–xiv, xvi–xviii, 7, 13–14, 17–18, 26–27, 30–31, 35, 38, 43–44, 46, 50, 52, 61, 70–75, 78–80, 82–85, 96, 112–14, 137, 183–84. *See also* telegraph lines; transmission lines
The Wire Tappers (Stringer). See Stringer, Arthur
wiretapping, 78–82, 173. *See also* Stringer, Arthur
Wright, Frank Lloyd, 141

Young, Louise B.: *Power Over People*, 144–47

CPSIA information can be obtained
at www.ICGtesting.com
Printed in the USA
LVHW011709120619
621007LV00005B/30